T0134825

Studies in Computational Intelligence

Volume 664

Series editor

Janusz Kacprzyk, Polish Academy of Sciences, Warsaw, Poland
e-mail: kacprzyk@ibspan.waw.pl

About this Series

The series "Studies in Computational Intelligence" (SCI) publishes new developments and advances in the various areas of computational intelligence—quickly and with a high quality. The intent is to cover the theory, applications, and design methods of computational intelligence, as embedded in the fields of engineering, computer science, physics and life sciences, as well as the methodologies behind them. The series contains monographs, lecture notes and edited volumes in computational intelligence spanning the areas of neural networks, connectionist systems, genetic algorithms, evolutionary computation, artificial intelligence, cellular automata, self-organizing systems, soft computing, fuzzy systems, and hybrid intelligent systems. Of particular value to both the contributors and the readership are the short publication timeframe and the worldwide distribution, which enable both wide and rapid dissemination of research output.

More information about this series at http://www.springer.com/series/7092

Nadia Nedjah · Heitor Silvério Lopes
Luiza de Macedo Mourelle
Editors

Designing with Computational Intelligence

 Springer

Editors
Nadia Nedjah
Department of Electronics Engineering
  and Telecommunications
State University of Rio de Janeiro
Rio de Janeiro
Brazil

Luiza de Macedo Mourelle
Department of Systems Engineering
  and Computer Science
State University of Rio de Janeiro
Rio de Janeiro
Brazil

Heitor Silvério Lopes
Department of Electronics
Technological Federal University of Paraná
  (UTFPR)
Curitiba, Paraná
Brazil

ISSN 1860-949X   ISSN 1860-9503   (electronic)
Studies in Computational Intelligence
ISBN 978-3-319-83124-4   ISBN 978-3-319-44735-3   (eBook)
DOI 10.1007/978-3-319-44735-3

Printed on acid-free paper

This Springer imprint is published by Springer Nature
The registered company is Springer International Publishing AG
The registered company address is: Gewerbestrasse 11, 6330 Cham, Switzerland

Preface

In recent years, computational intelligence has attracted many researchers' attention and so became a consolidated methodology to automatically create new competitive solution to complex real-world problems. Concise and efficient synthesis of a variety of systems has been generated using computationally intelligent techniques. This book puts together a set of chapters, in which some real-world applications of interest are approached using computational intelligence. In the following, we give a brief description of the main contribution of each of the included chapters.

In Chap. 1, which is entitled *On Using Fuzzy Logic to Control a Simulated Hexacopter Carrying an Attached Pendulum*, the authors propose an approach based on multiple interconnected fuzzy controllers, aiming at controlling the various aspects related to maneuverability of a hexacopter carrying a free payload forming a pendulum. They simulated the behavior produced by the proposed control system on a robotics simulation environment and analyzed the achieved results in terms of flight stability, roll, pitch and yaw movements. The authors claim that the results show the feasibility of the proposed approach, which allowed the flight stability of the hexacopter.

In Chap. 2, which is entitled *Monocular Pose Estimation for an Unmanned Aerial Vehicle Using Spectral Features*, the authors propose a visual position and orientation estimation algorithm based on the discrete homography constraint, induced by the presence of planar scenes and the so-called spectral features in the image. The authors claim the their approach has some unique characteristics, which are the selection of an appropriate distribution of the features, no requirement of an initialization step nor a search for features and no impact of the presence of corner-like features in the scene. The authors tested the proposed pose estimation algorithm in a simulated dataset. They prove the robustness of the spectral features in different conditions using a conveyor belt.

In Chap. 3, which is entitled *Simultaneous Navigation and Mapping in an Autonomous Vehicle Based on Fuzzy Logic*, the authors present a navigation control and mapping of an autonomous car using fuzzy logic, enabling automatic obstacle avoidance in unknown environments. The author's strategy is based on a map of the environment to plan the trajectories avoiding obstacles through the search algorithm

A*. They evaluated the proposed approach in a virtual environment, where the autonomous car moves among different obstacles.

In Chap. 4, which is entitled *Fully Scalable Parallel Hardware for Wheeled Robot Navigation Using Fuzzy Control*, the authors describe a reconfigurable-efficient architecture for fuzzy controllers, suitable for embedding in final products. They show that the architecture is parameterizable allowing the setup and config-uration of the controller so it can be used as a control in many applications. The authors present and evaluate an application of the fuzzy controller hardware architecture in the supervision of a wheeled robot during navigation in an unknown environment.

In Chap. 5, which is entitled *Nonlinear Correction for an Energy Estimator Operating at Severe Pile-Up Conditions*, the authors describe how computational intelligence can be used to assist during energy estimation performed by an optimal linear method. They use an artificial neural network that is trained aiming at cor-recting the nonlinearities introduced by the signal pile-up statistics. The authors evaluate the efficiency of the various energy estimation methods using simulation data under various signal pile-up scenarios.

In Chap. 6, which is entitled *Non-supervised Learning Applied to Analysis of Topological Metrics of Optical Networks*, the authors offer a systematic method to analyze different backbone optical networks, based on a non-supervised algorithm for clustering. They investigate the power of a recently proposed topological metrics, which along with three others are applied to identify the best canonical model to represent real backbone optical networks. The authors claim that according to the obtained results, the proposed clustering procedure that the investigate metric is the best metric to explain the installed capacity for the analyzed networks.

In Chap. 7, which is entitled *Mole Features Extraction for a Melanoma Recognition System*, the authors propose three algorithms to extract features of skin moles based on dermatological studies, using digital image processing techniques existing in the lecture. They also evaluate these features as input to classifiers creating a melanoma recognition, and indicating whether it is a melanoma or normal mole. The authors analyze the obtained results, which are shown through ROC curve and 10-fold cross-validation from two dermatological datasets Atlas of Clinical Dermatology and DermNet NZ.

In Chap. 8, which is entitled *Human–Machine Musical Composition in Real-Time Based on Emotions Through a Fuzzy Logic Approach*, the authors present a method for representing human emotions in the context of musical composition, which is used to artificially generate musical melodies using fuzzy logic. They tested the generated melodies with listeners in an experiment aiming at of verifying if these melodies can produce emotions in them and whether those emotions match the emotional intentions captured by humans.

In Chap. 9, which is entitled *A Recursive Genetic Algorithm-Based Approach for Educational Timetabling Problems*, the authors address the educational timetabling problem for multiple courses, aiming at finding solutions that satisfy the hard constraints and minimize the soft constraint violations. They propose a simple, scalable and parameterized recursive approach to solve timetabling problems for multiple courses with genetic algorithms, which are efficient search methods used to achieve an pseudo-optimal solution.

In Chap. 10, which is entitled *Evolving Connection Weights of Artificial Neural Network Using a Multi-Objective Approach with Application to Class Prediction*, the authors investigate the applicability of two novel multi-objective evolutionary algorithms: Speed constrained multi-objective particle swarm optimization and multi-objective differential evolution algorithm based on decomposition with dynamical resource allocation. They compare the obtained results using the hypervolume as quality indicator.

In Chap. 11, which is entitled *Diversification Strategies in Evolutionary Algorithms: Application to the Scheduling of Power Network Outages*, the authors propose different strategies to avoid and/or fix premature convergence of evolutionary algorithms. They claim that high diversification level is maintained throughout the evolution process, so that an adequate trade-off between solution quality and computational cost is achieved. Through numerical results, they illustrate the application of the proposed strategies and respective impact on the quality and computational cost of solutions.

In Chap. 12, which is entitled *WBdetect: Particle Swarm Optimization for Segmenting Weld Beads in Radiographic Images*, the authors present an approach for automatically segmenting weld beads in double wall double image X-ray photographs by combining two known methods: Particle swarm optimization and dynamic time warping. They show through experiments that the achieved results are promising and outperform existing approach.

The editors are very much grateful to the authors of this volume and to the reviewers for their tremendous service by critically reviewing the chapters. The editors would like also to thank Prof. Janusz Kacprzyk, the editor-in-chief of the Studies in Computational Intelligence Book Series and Dr. Thomas Ditzinger, Springer Verlag, Germany for the editorial assistance and excellent collaboration to produce this important scientific work. We hope that the reader will share our excitement to present this volume and will find it useful.

Rio de Janeiro, Brazil   Nadia Nedjah
Curitiba, Brazil   Heitor Silvério Lopes
Rio de Janeiro, Brazil   Luiza de Macedo Mourelle
May 2016

Contents

List of Figures

List of Tables

Chapter 1
On Using Fuzzy Logic to Control a Simulated Hexacopter Carrying an Attached Pendulum

**Emanoel Koslosky, Marco A. Wehrmeister, João A. Fabro
and André S. de Oliveira**

Fuzzy logic is used in many applications from industrial process control to automotive applications, including consumers trend forecast, aircraft maneuvering control and others. Considering the increased interest in using of multi-rotor aircrafts (usually called drones) for many kinds of applications, it is important to study new methods to improve multi-rotor maneuverability while controlling its stability in a proper way. Controlling the flight of multi-rotors, specially those equipped six rotors, is not a trivial task. When considering the design of such a control systems, traditional approaches such as PD/PID are very difficult to design, in spite of being easily implementable. This work proposes an approach based on multiple interconnected fuzzy controllers, aiming to control the various aspects related to maneuverability of a hexacopter carrying a free payload forming a pendulum. The behavior produced by such a control system has been simulated on a well-known robotics simulation environment and analyzed in terms of flight stability, as well as roll, pitch and yaw movements. The results show the feasibility of the proposed approach in keeping the hexacopter flying in a stable way.

E. Koslosky (✉) · M.A. Wehrmeister · J.A. Fabro · A.S. de Oliveira
Federal University of Technology - Paraná (UTFPR), Curitiba, Brazil
e-mail: ekosky@gmail.com

M.A. Wehrmeister
e-mail: wehrmeister@utfpr.edu.br

J.A. Fabro
e-mail: fabro@utfpr.edu.br

A.S. de Oliveira
e-mail: andreoliveira@utfpr.edu.br

© Springer International Publishing Switzerland 2017
N. Nedjah et al. (eds.), *Designing with Computational Intelligence*,
Studies in Computational Intelligence 664,
DOI 10.1007/978-3-319-44735-3_1

1

1.1 Introduction

Nowadays, technology advances and cost reductions have popularized the use of small electromechanical aircrafts in many distinct application fields, such as video recording, plantation inspections, search-and-rescue assistance, military and civil surveillance applications, among others. Multi-rotor helicopters (also known as drones) are among the popular small electromechanical aircrafts that are being used in such applications. Moreover, some of these new applications demand multi-rotor helicopters that fly autonomously, as presented in [4, 7]. Thus, the multi-rotor helicopter must have additional computational systems on top of the more basic movement and stabilization control systems. These computational systems provide higher level capabilities to support the mission accomplishment. Therefore, Unmanned Aerial Vehicles (UAV) are the preferred choice for these applications, due to the cost reductions obtained from eliminating the need of high-skilled and trained pilots.

There are several topologies for multi-rotor helicopter, varying on the number of rotors (i.e. motor and propeller), as well as on the position of these rotors onto the aircraft frame. The most common multi-rotor helicopter has 4 rotors and is called quadcopter. However, recently, other multi-rotor helicopter topologies are becoming popular, such as those with 6 rotors, the so-called hexacopter, as discussed in [9].

The UAV stabilization is commonly performed by hybrid control approaches (parallel, cascade) with multiple PID controllers, like works of [1, 2]. However, these methods require a precise mathematical formulation or identification of UAV dynamics to minimize the disturbance and stabilize the system, as discussed in [11].

Adaptive algorithms can be applied to establish multivariable systems (like UAVs) with more efficiency which classical strategies. In [5] is discusses a approach based on artificial neural networks to trajectory control of UAVs. In [6] the genetic algorithm is applied to establish a hexacopter. In [3] a fuzzy logic method is used to position control of a hexacopter. However, the main focus of previous works is the UAV stabilization over linear disturbances and is not evaluate the proposed control strategies over nonlinear disturbances, like a variable payload.

This work focuses on the control system for the movement and stabilization of a Hexacopter, whose rotors have been configured as a "Hexa +" topology. A multi-layer controller has been proposed and integrates multiple fuzzy controllers. The outputs from these fuzzy controllers must be applied on each rotor accordingly, in order to get the correct Hexacopter movements. A closed control loop is obtained by reading of sensors that measure the position and the movements of the Hexacopter, which, on the other hand, are used as feedback information to the proposed multilayer controller. The main goal is to create a robust and flexible controller that is able to keep the Hexacopter stability when moving or hovering, even when it carries a free or loose payload that changes its center of gravity.

The major challenge tackled in this work is the switching among fuzzy controllers at the right moment. For instance, lets assume that a hexacopter starts on the ground and receives a command to fly to a certain position, e.g. 2 m in latitude, 5 m in longitude and 3 m upward. To achieve the commanded position, every movement must be executed properly, and hence, all controllers must work cooperatively to achieve the goal. In other words, the fuzzy controller that controls the longitudinal position cannot override the other controllers actions. On the other hand, when the fuzzy controllers do not work well together, a controller outputs can take the rotors actuation over the outputs of the other controllers. In this case, there must exist a high-level controller that performs the contention of misbehaved controllers.

The proposed approach has been validated through simulation. For that, a model of a hexacopter has been created in the V-REP simulation environment. A free payload has been attached to the hexacopter forming a pendulum. Thus, the proposed multi-layer controller must control the hexacopter movement when it is commanded to move to another position, while keeping the its body stabilized during the flight. Results indicate that the proposed approach is robust since it allows the hexacopter move from one position to another, even though it must carry a moving payload.

The reminder of this chapter is organized as follows: Sect. 1.2 provides an overview of the control problem; Sect. 1.3 presents the proposed stabilization and movement multi-layer fuzzy controller; Sect. 1.4 provides details on each fuzzy controller that comprises the proposed controller; Sect. 1.5 discusses the conducted experiments and the obtained results; finally, Sect. 1.6 draws some conclusions and presents future work directions.

1.2 Description of the Controlled Plant: The Hexacopter

This section describes briefly the system under control, i.e. the hexacopter, as *a control plant*. Figure 1.1 shows the hexacopter, which is composed by six rotors organized as "Hexa +" topology.[1]

By activating the rotors accordingly, it is possible to control the hexacopter maneuvering through the X, Y and Z axes.

In general, each plant must be analyzed to discover the interaction of each force. In this case, the thrust force produced by speeding up or slowing down some rotors leads the hexacopter toward the desired direction (on each axis), i.e. the vectors of the forces acting on the plant.

To understand the movements performed by the hexacopter it is worth to take a look at the forces acting in frame.

In the Figs. 1.2 and 1.3 show the forces the rotors imposed to the frame. If these forces are unbalanced the hexacopter start a rotation around the Y-axis and therefore this rotation makes the hexacopter to start a movement over the X-axis. The hexacopter moves forward (Fig. 1.2) if the rear rotor has a value greater than the front rotor. If the front rotor has a value greater than the rear rotor, the hexacopter moves backward or, if it is going forward in this situation, this rotation makes it to slows down.

---

[1] See *APM:Coter – Connect ESCs and Motors*, http://copter.ardupilot.com/wiki/connect-escs-and-motors/

Fig. 1.1 The plant to be controlled: (i) X axis angle is the Roll rotation, (ii) the Y axis is the Pitch rotation, and (iii) the Z axis is the Yaw rotation. The *arrow* direction means positive

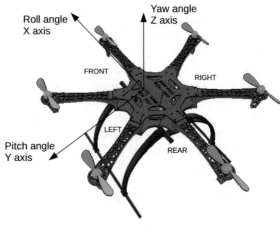

Fig. 1.2 The rear rotor has greater value than front rotor, the hexacopter moves forward

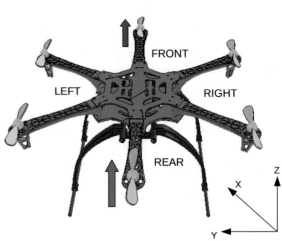

Fig. 1.3 The front rotor has greater value than rear rotor, the hexacopter moves backward

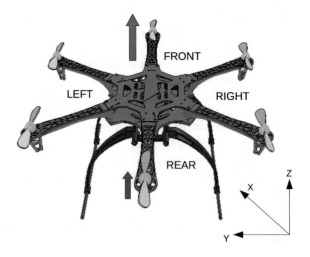

Fig. 1.4 The *left side* rotors have value greater than applied on the *right side* rotors, the hexacopter moves to the *right*, or slows down if it is moving to the *left*

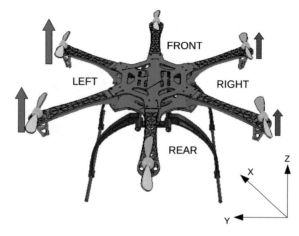

Fig. 1.5 The *right side* rotors have value greater than applied on the *left side* rotors, the hexacopter moves to the *left*, or slows down if it is moving to the *right*

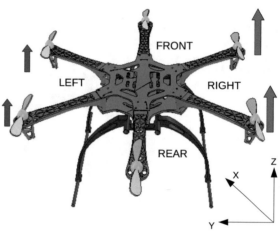

The maneuvers to the right and to the left are achieved by applying forces on the side rotors with different proportion. Thus, it makes the hexacopter rotate around the X-axis and a movement over the Y-axis occurs. For moving to the right or to slow down the movement to the left, the left side rotors have value greater than applied on the right side rotors, as shown the Fig. 1.4. For moving to left or to slow down the movement to right, these forces are inversely applied between the left and right side rotors as shown in the Fig. 1.5.

To rotate the hexacopter in the Z-axis (the yaw movement) the forces are applied alternating among the rotors, shown in Figs. 1.6 and 1.7. It is important to note that if a rotor is set to rotate clockwise, therefore, the adjacent rotor is set to rotate counter-clockwise. The real propellers are built with clockwise twist and counterclockwise. If the forces are applied with some difference between the adjacent rotors, a gyroscopic effect begin to act on the frame. This effect is used to rotate the hexacopter around the Z-axis.

Fig. 1.6 The different forces applied to the adjacent rotors to rotate the hexacopter clockwise

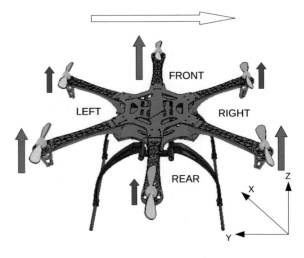

Fig. 1.7 The different forces applied to the adjacent rotors to rotate the hexacopter counter-clockwise

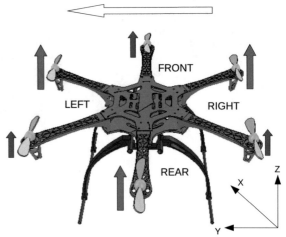

Finally, If all rotors receive the same force, the hexacopter could lift off, land or keep hovering, depending of the intensity. The Figs. 1.8 and 1.9 shows this effect. The hexacopter goes up when the force is high and it goes down when the force is low.

1.3 Proposed Multi-layer Controller

The proposed controller implements a closed loop that comprises the three layers. Data produced as output in one layer is passed as input to the next layer.

Fig. 1.8 With high force,
the hexacopter goes up

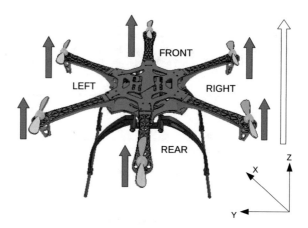

Fig. 1.9 With low force, the
hexacopter goes down

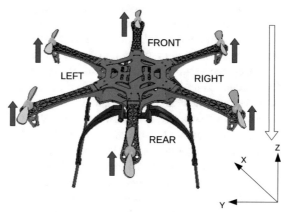

The proposed multi-layer fuzzy controller is based on [12] and is depicted in Fig. 1.10. The *Control box* is composed by a pre-processing phase (first layer), a set of fuzzy controllers (second layer), and post-processing phase (third layer).

As one can observe, after the post-processing phase, the control outputs are applied onto the plant by means of the hexacopter rotors that actuate on the hexacopter movement and stabilization. The sensors perceive the changes on the plant controlled variables, and hence, provide the feedback to the controller. The controller, in turn, compares these input values with the reference values established as setpoints thereby closing the control loop [8].

The pre-processing phase (first layer) is responsible for acquiring data from the input sensors, process the input movement commands, as well as calculate the controlled data used as input to the fuzzy controllers in second layer. Before the multi-layer controller starts its execution, there is an initialization phase that is performed within the first layer. The target position is set as the current position, so that the hexacopter does not move before receiving any command. Gyroscope and accelerometer sensors are calibrated and the GPS sensor is initialized by gathering at least four

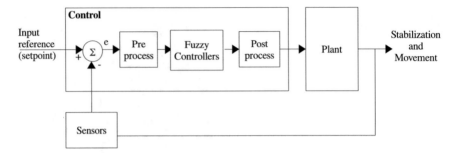

Fig. 1.10 Inside the control box. Composed by preprocess, fuzzy controllers and postprocess

satellites. During the execution phase, the first layer is responsible to calculate the input variables to the fuzzy controllers: (i) the angular and linear distance (delta error) for X, Y, and Z axes between the current hexacopter position and the target position; (ii) the rotation and translation movement matrices to translate 3 axes movement into the speed related to the ground (i.e. X and Y axis). In addition, it is responsible to convert the input movement commands into setpoints for X, Y and Z positions. Movements commands are composed of three values representing the positive or negative movement along X, Y and Z axes related to the current positions, i.e. a command indicates a relative position. Thus, when a new command is received, the first layer will convert it to a absolute position. Then, when the control system is executing, this layer uses the GPS coordinates to determine the error in the distance from the hexacopter to the target position. These calculated errors in position are the inputs to the fuzzy controllers (Euler X, Euler Y and Euler Z errors).

The second layer contains five fuzzy controllers, which act on issues regarding the hexacopter movement, namely hovering stabilization, vertical and horizontal movement and heading. As mentioned, these controllers take as input the data produced in the first layer and generate output for the third layer. The generated outputs represent the actuation on the six rotors for performing pitch, roll, yaw moves for all maneuvers necessary to reach the target position. The fuzzy controllers are discussed in details in the next section.

The post-processing phase (third layer) is responsible for coordinating the fuzzy controllers outputs. As mentioned, in order to perform a proper maneuver, the proposed multi-layer controller establishes a priority on movements needed to complete a maneuver. When a new command is received, i.e. a new target point is set, the hexacopter must firstly reach the target altitude. Then, the hexacopter must turn until its front aims the target position. Finally, the hexacopter moves horizontally towards the target position. This layers also performs a threshold limits control by means of output values saturation, in order to keep the hexacopter stability while flying or hovering.

1.3.1 Pre-processing Phase

In the pre-processing phase some calculations are applied onto the fuzzy controllers. Each fuzzy controller (see Sect. 1.4) has at least two inputs [12]: the error e(t) and some derivative variable such as speed.

The error is calculated according to Eq. 1.1. In other words, e(t) is the difference between the reference value r(t) and the current value of an sensor y(t).

The reference is a setpoint established by an operator or other controller. Some inputs have the reference always set to zero so that the error e(t) is the opposite value of sensor, i.e. -y(t).

$$e(t) = r(t) - y(t) \tag{1.1}$$

The errors calculated in the pre-processing phase are related to the following variables. The rotation around the three axes is the Euler angle (in radians) and it is measured by a gyroscope. The following outputs are generated:

- Angle on X axis: Roll_error
- Angle on Y axis: Pitch_error
- Angle on Z axis: Yaw_error

The *distance* to the target position (i.e. the error between the current hexacopter position and the setpoint on the three axes) is measured in meters using GPS and calculated as described bellow.

The following outputs are generated:

- Horizontal distance on X axis: distX_error
- Horizontal distance on Y axis: distY_error
- Vertical distance on Z axis: distZ_error

The distance ds is calculated to determine the distance to the target position that is decomposed by X and Y axes, as shown in Eqs. 1.2 and 1.3.

Then, the Euclidean Distance is calculated to obtain the real distance d to the target (see Eq. 1.4). Euclidean Distance d is also used in Eq. 1.6 to calculate the speed v(t).

$$\Delta \, ds_x = s_{x_t} - s_{x_{(t-1)}} \tag{1.2}$$

$$\Delta \, ds_y = s_{y_t} - s_{y_{(t-1)}} \tag{1.3}$$

$$\Delta \, d = \sqrt{\Delta d_x^2 + \Delta d_Y^2} \tag{1.4}$$

Moreover, the angle of the movement is determined by the arctangent as shown in Eq. 1.5. The angle a(t) is used as the new yaw setpoint.

$$a(t) = arctangent(\frac{\Delta d_x}{\Delta d_y}) \tag{1.5}$$

The *linear speed* is measured in meters per second and is calculated according to Eq. 1.6 as discussed bellow. The following outputs are generated:

- Horizontal speed on X axis: SpeedX_error
- Horizontal speed on Y axis: SpeedY_error
- Vertical speed on Z axis: SpeedZ_error

The distance ds and the time interval Δt are used to determine the current speed of the hexacopter.

The time interval Δt is obtained by measuring the time instant on which two consecutive values of Euclidian Distance are calculated.

$$v(t) = \frac{\Delta d}{\Delta t} \tag{1.6}$$

Despite of acceleration is a derivative of speed, such a variable is taken directly from accelerometer sensor. The acceleration information is used together other others measurements like Euler angles in order to avoid oscillation movements. For instance, the acceleration measured over Y-axes is used together Euler angle error measured around the X-axis. If the Euler angle erro is zero, it means the hexacopter is stabilized accordingly to the X-angle. But it could be moving over the in Y-axis like drifting. The controller must to slow down this movement. If not stop this drifting, a oscillation begin appear. In order to the fuzzy controller realize this moment, the acceleration measurement is used as input together with Euler angle to be processed by controller. The controllers are described in detail in the Sect. 1.4.

Another calculation present in pre-processing phase is the rotation matrix [15]. It is used to obtain the speed and distance error over the X and Y axes related the pose of hexacopter and the inertial frame. In other words, the information from GPS tells the hexacopter position on the world (the inertial frame) but nothing about the pose of it. By applying the rotation matrix calculation, using the Euler Z-angle error, it is possible to discover the pose of hexacopter on the world as well as the speed related to its X and Y axis. With this information the controller can determine the correct forces to be applied onto each rotor.

1.3.2 Post-processing Phase

The post-processing phase determines the movement sequencing, saturation and so on. The output of fuzzy controllers are provided as the inputs to this phase. Such a behavior is explained as the Finite State Machine (FSM) depicted in Fig. 1.11.

There are many approaches for controlling the movement of a hexacopter. For instance, one could create an algorithm in which the hexacopter flies directly to the target position in the three-dimensional space by changing the vehicle altitude and horizontal position at same time.

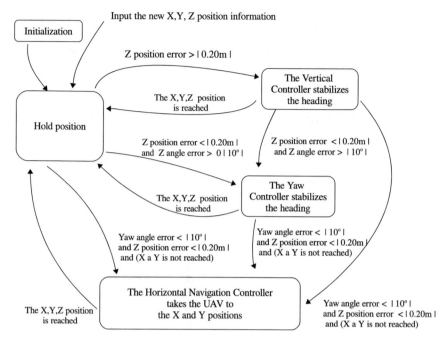

Fig. 1.11 FSM for sequencing the hexacopter maneuver process. (X, Y, Z) inputs represent the new target position of the hexacopter

Due to the instability caused by the loose payload, the controller proposed in this work adopts an approach that sequentializes the flying movements. As presented in Fig. 1.11, the first movement of hexacopter is to reach the desired altitude (Z-axis position). Thereafter, the controller commands the hexacopter to aim directly at the target position by rotating on the Y-axis. Finally, the controller commands the hexacopter to move along with X and Y axes, in order to reach the target position.

After the initialization, the systems goes to the *Hold Position* state. In this state, *Roll*, *Pitch* and *Hovering* movements are stabilized by their fuzzy controllers. Once a new command is provided by the operator, it changes the X, Y and Z position setpoints, leading to an increase on error values as described in Sect. 1.3.1. The proper fuzzy controller is activated when the threshold of one of its input values is reached. For instance, *Horizontal Navigation* fuzzy controller starts when the altitude and yaw is under a certain threshold. Such thresholds controls the hexacopter stabilization. It is important to note that if some external disturbance interfere with the hexacopter stabilization, the controller stops the horizontal movement until the input values reach their thresholds.

The proposed multi-layer controller implements the saturation control in the post-processing phase, in order to avoid an individual fuzzy controller to override other fuzzy controllers outputs by means of dominating the actuation on the plant.

Therefore, once the output of logic control is calculated, it is important to determine the exact value that must be applied on each rotor.

When throttle value is the same to all rotors, the hexacopter keeps hovering on the same position. On the other hand, the output from roll and pitch stabilization are applied proportionally as a gain to the rotors throttle according to the Eqs. 1.7–1.12.

$$PropellerForceFRONT = Othrottle - Othrottle \times Opitch; \tag{1.7}$$

$$PropellerForceRIGHT\_FRONT = Othrottle - Othrottle \times (Oroll/2); \tag{1.8}$$

$$PropellerForceRIGHT\_REAR = Othrottle - Othrottle \times (Oroll/2); \tag{1.9}$$

$$PropellerForceREAR = Othrottle + Othrottle \times Opitch; \tag{1.10}$$

$$PropellerForceLEFT\_REAR = Othrottle + Othrottle \times (Oroll/2); \tag{1.11}$$

$$PropellerForceLEFT\_FRONT = Othrottle + Othrottle \times (Oroll/2); \tag{1.12}$$

In these equations, `PropellerForce<Rotor Position>` is the value applied on the rotor, the `Othrottle` is the output from *Hovering fuzzy controller*, and the `Optitch` and `Oroll` are the outputs from, respectively, *Pitch* and *Roll fuzzy controllers*. It is worth mentioning that: (i) to maintain the opposite feedback into the mesh, the output value is obtained by subtracting `Oroll` from the throttle value for rotors at the right side of the hexacopter, as well as by subtracting `Opitch` for the front rotor. Similarly, for the left-side and rear rotors, respectively, `Oroll` and `Opitch` are added to the throttle value; and (ii) `Oroll` values are proportional to the amount of rotors on the right/left sides of the hexacopter, i.e. `Oroll` value is divided by two. Such proportional values avoid that right and left `Othrottle` values do not override the front and rear `Othrottle` values.

1.4   The Fuzzy Controllers

1.4.1   The Fuzzy Method

A fuzzy controller can be created with a variety of types of membership functions such as trapezoidal, triangle, Gaussian bell curve function, and others. In addition, these function may be of receive many inputs and provide a simple output (MISO) or receive many inputs and provide many outputs (MIMO).

The fuzzy controller proposed in this work are composed five independent fuzzy controllers. These controllers are built from MISO membership functions defined as trapezoidal and triangle forms. The `min()` operator has been used in the rule inferences and the result is done by `max()` operator.

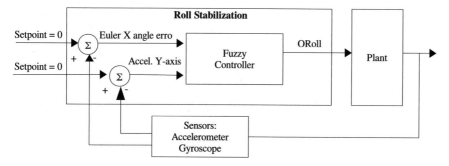

Fig. 1.12 The roll stabilization fuzzy controller

Moreover, the defuzzification method used in this work is Center of Gravity (COG). The next sections provides details on these five independent fuzzy controller.

1.4.2 Roll Stabilization

Roll is the movement obtained through the rotation around the X-axis, i.e. front-to-back axis. The fuzzy controller named Roll Stabilization controls the stabilization of the hexacopter while it is performing the roll maneuver. Figure 1.12 shows the block diagram of this controller.

This controller has two input data. The first input is error in roll angle Euler approximation. The roll angle is calculated through the Euler approximation of the current X angle and the target X-axis angle. Figure 1.13 shows the linguistic variable membership function representing the fuzzification of the error in the roll angle Euler approximation. The second input is the perceived movement in Y-axis represented as the acceleration in Y-axis obtained from the accelerometer over the time. Figure 1.14 shows the linguistic variables and the membership functions representing

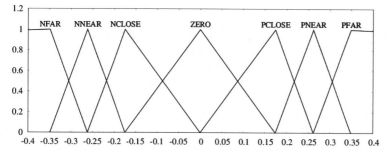

Fig. 1.13 Input linguistic variables and their membership functions for the roll angle, Euler approximation error of X-axis angle

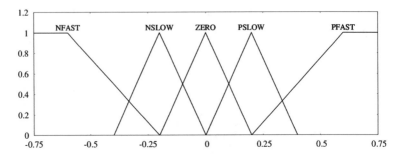

Fig. 1.14 Input linguistic variables and their membership functions for: Y-axis accelerations

Table 1.1 Control rules for roll stabilization

Roll angle AccelY	NFAR	NNEAR	NCLOSE	ZERO	PCLOSE	PNEAR	PFAR
NFAST	PMAX	PMIN	ZERO	PMIN	NMIN	NMID	NMAX
NSLOW	PMAX	PMIN	ZERO	ZERO	NMIN	NMID	NMAX
ZERO	PMAX	PMID	PMIN	ZERO	NMIN	NMID	NMAX
PSLOW	PMAX	PMID	PMIN	ZERO	ZERO	NMIN	NMAX
PFAST	PMAX	PMID	PMIN	NMIN	ZERO	NMIN	NMAX

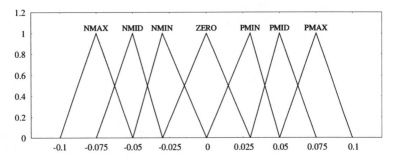

Fig. 1.15 Output linguistic variables and their membership functions for: ORoll

the fuzzification of Y-axis acceleration. Its worth mentioning that the "N" and "P" prefixes of variables names stand for, respectively, Negative and Positive.

The roll stabilization fuzzy controller is composed of 35 rules as show in Table 1.1. The output of this controller is the omega roll variable (ORoll), whose values are depicted in Fig. 1.15. The defuzzification of ORoll variable creates the values that control the rotation speed of right- and left-hand side rotors, which, in turn, produce enough force to make the hexacopter rotate in the X-axis. It is important to highlight that X-axis and Y-axis acceleration and also roll and pitch angle error variables are used to minimize (or correct) the stabilization interference caused by pendulum effect created by the free payload.

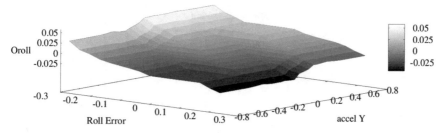

Fig. 1.16 Fuzzy control surface for roll stability control

The fuzzy control surface graphic is shown in Fig. 1.16; it depicts the relationship among inputs and output. It is possible to note that at near the point zero of roll error the output is smooth. When the the roll error is far from zero the correction is greater. Also we can see that acceleration over the Y-axis in all the surface plays a important role: it avoids the oscillation. For instance, suppose the error is about 0.1, a rotation to the right makes the hexacopter to fly to the right direction. When the hexacopter is rotating and the angle error is reaching zero, the hexacopter starts to fly faster. Without considering the information on acceleration, the hexacopter tends to oscillate. Another way to see this effect is the following: the acceleration over Y axis is negative when roll error is positive and vice-versa. However, there are situations in which the hexacopter moves in positive direction in respect to the Y-axis, the roll error is positive. This means that the hexacopter must start to fly slower and the angle must be kept unchanged or raised. Without the acceleration information, the output of this controller could become zero (when roll error is zero) and the hexacopter might drift. Some derivative value in time, e.g. speed or acceleration, helps to avoid oscillation or sliding.

1.4.3 Pitch Stabilization

Pitch is the movement obtained through the rotation around the Y-axis, i.e. side-to-side axis. The pitch stabilization fuzzy controller holds the hexacopter stabilization while it is performing the pitch maneuver. Figure 1.17 shows the block diagram of this controller.

This controller is very similar to the Roll Stabilization controller. It has two input variables: (i) the error in pitch angle calculated through Euler approximations; and (ii) X-axis acceleration to indicate a possible movement in the X-axis. The linguistic variables and the membership function values for these two input variables are depicted in Figs. 1.18 and 1.19, respectively. In the same way, pitch stabilization is also defined by five rules as described in Table 1.2, and has only one output the omega pitch variable (OPitch) depicted in Fig. 1.20. The definition of these variables and rules are exactly the same as in the Roll Stabilization fuzzy controller. However,

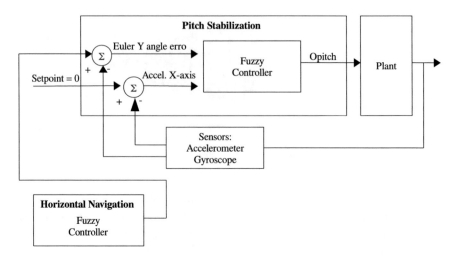

Fig. 1.17 The pitch stabilization fuzzy controller

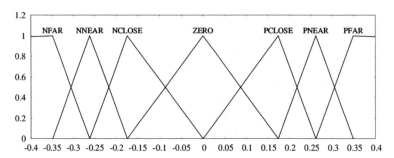

Fig. 1.18 Input linguistic variables and their membership functions for: **Pitch angle**, Euler approximation error of Y-axis angle

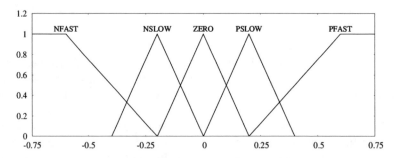

Fig. 1.19 Input linguistic variables and their membership functions for: X-axis

Table 1.2 Control rules for pitch stabilization

Roll angle	NFAR	NNEAR	NCLOSE	ZERO	PCLOSE	PNEAR	PFAR
AccelY							
NFAST	PMAX	PMIN	ZERO	PMIN	NMIN	NMID	NMAX
NSLOW	PMAX	PMIN	ZERO	ZERO	NMIN	NMID	NMAX
ZERO	PMAX	PMID	PMIN	ZERO	NMIN	NMID	NMAX
PSLOW	PMAX	PMID	PMIN	ZERO	ZERO	NMIN	NMAX
PFAST	PMAX	PMID	PMIN	NMIN	ZERO	NMIN	NMAX

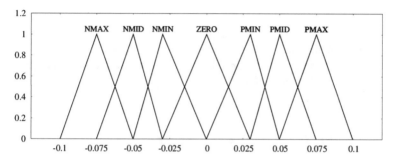

Fig. 1.20 Output linguistic variables and their membership functions for: OPitch

it is important to highlight that the defuzzification process on OPitch generates values that control the speed of the front and read rotors making the hexacopter rotate on the Y-axis. It is done by Horizontal navigation controller described later.

Similarly, the pendulum effect mentioned previously is minimized or mitigated by means of using the pitch angle error and Y-axis acceleration variables. The roll and pitch stabilization fuzzy controllers are key to maintain the hexacopter stabilized while it is flying or hovering. Equally important is the way these controllers actuate on the rotors. The correction applied to a controlled "defuzzified" variable is proportional to its current value, see Eqs. 1.5–1.11. In other words, instead of simply summing a new absolute actuation value to a variable (e.g. the rotor speed), the gain is proportional to current value. This improves the controller efficiency in extreme situations, e.g. when the corrective value is insignificant (compared to the current value) or the correction value is too high, avoiding aggressive corrections, and hence, improving stability. Finally, it is worth noting that the overall hexacopter stabilization is performed by both roll and pitch stabilization fuzzy controllers.

The surface graphic about pitch stability control depicted in Fig. 1.21 shows the relationship between pitch angle error and the accelerometer information over the X-axis. Both input values work together to stabilize the hexacopter over the Y-axes and avoid oscillation over the X-axis. For instance. Suppose the pitch controller receive the pitch error as input value close to zero and as the acceleration input value high. It means the hexacopter is in movement forward even though the pitch angle error is zero. Thus, the controller must to slow down it. The control surface shows

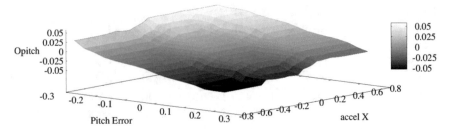

Fig. 1.21 Fuzzy control surface for pitch stability control

that the output pitch angle (Opitch) is greater than zero in this situation, it makes the hexacopter rotate around Y-axis counter-clockwise causing it to slow down.

The forward movement is obtained by changing the Y-axis setpoint computed by *Horizontal navigation controller*.

1.4.4 Heading to Goal (Yaw Controller)

Yaw is the rotation movement around the vertical axis, that is, the Z-axis. The yaw controller (or heading goal controller) is responsible for pointing the hexacopter front to the target position, keeping this position until it arrives at the destination. Figure 1.22 shows the block diagram of this controller.

To allow this controller to perform such a task, two input variables are needed. The yaw angle is similar to roll and pitch angles, and hence, are calculated based on the Euler approximation of Z-axis angle between the current angle position of the hexacopter front and the target position. Figure 1.23 shows the linguistic variable and the membership function values of yaw angle. The distance to the goal is the second input data utilized. It indicates how far the hexacopter from the target position. The

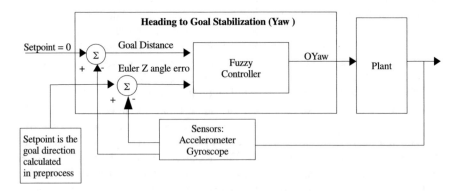

Fig. 1.22 The heading to goal stabilization fuzzy controller (The Yaw Stabilization)

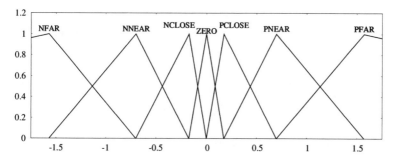

Fig. 1.23 Input linguistic variables and their membership functions for: Yaw angle error

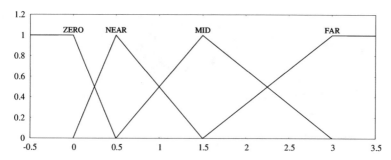

Fig. 1.24 Input linguistic variables and their membership functions for: Goal distance to determines the hexacopter heading

Table 1.3 Yaw control rules

Yaw angle	NFAR	NNEAR	NCLOSE	ZERO	PCLOSE	PNEAR	PFAR
Distance							
ZERO	ZERO	ZERO	ZERO	ZERO	ZERO	ZERO	ZERO
NEAR	PMAX	PMID	PMIN	ZERO	NMIN	NMID	NMAX
MID	PMAX	PMID	PMIN	ZERO	NMIN	NMID	NMAX
FAR	PMAX	PMID	PMIN	ZERO	NMIN	NMID	NMAX

goal distance is calculated from X- and Y-axis positions of the hexacopter and results in a polar coordinate indicating the angle and the distance to the target point. Goal distance linguist variable is depicted in Fig. 1.24.

This fuzzy controller has 28 rules as shown in Table 1.3.

These rules define the value of the output variable omega yaw (OYaw), presented in Fig. 1.25. OYaw is defuzzified and create actuation values that are applied on all rotors or on only a few of them. Depending on which rotor are affected the hexacopter turns clockwise or counter-clockwise. On the other hand, heading goal (yaw) controller gains priority over the other controllers when the hexacopter reaches

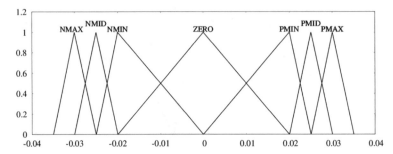

Fig. 1.25 Output linguistic variables and their membership functions for: OYaw

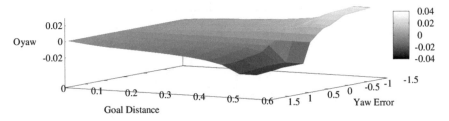

Fig. 1.26 Fuzzy control surface for Yaw stability control

a region within 70 cm radius around the target point. Thus, rover and pitch controllers are only responsible to main the hexacopter hovering stable on the target position.

Figure 1.26 shows the surface graphic of the *Yaw controller*. It works only when the goal distance is greater than the threshold, about 40 cm or higher in the *Goal Distance* anxis on the surface graphic. If the hexacopter is on the close to the target, the Yaw controller must be disabled. Otherwise, he stay spinning indefinitely.

1.4.5   The Horizontal Navigation

Horizontal navigation fuzzy controller controls the hexacopter fly on the X- and Y-axis. Figure 1.27 shows the block diagram of this controller. It takes as input the goal distance (see Sect. 1.4.4) and the horizontal speed.

The latter is calculated as traveled distance divided by time, i.e. the difference between the goal distance of two consecutive polar coordinates divided by the time elapsed between their calculation. Figure 1.28 shows the linguistic variable associated with the horizontal speed, while Fig. 1.29 shows the linguistic variable for the goal distance.

The rules of the horizontal navigation fuzzy controller are presented in Table 1.4. This controller results in the pitch angle for navigation (OPitchNavigation) as depict in Fig. 1.30 which, in turn, affects the pitch stabilization fuzzy controller. Specifically, the pitch angle for navigation moves the stability point towards the target

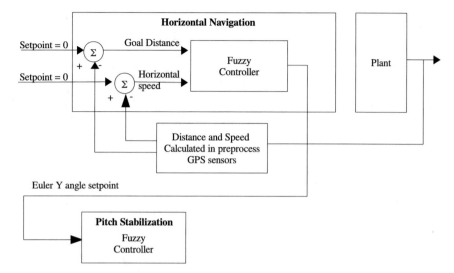

Fig. 1.27 The horizontal navigation fuzzy controller

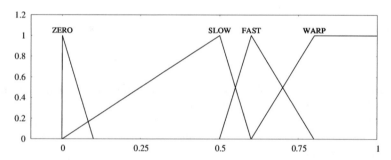

Fig. 1.28 Input linguistic variables and their membership functions for: Horizontal speed

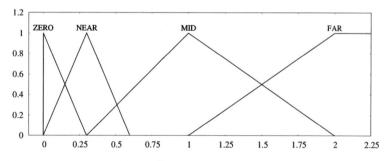

Fig. 1.29 Input linguistic variables and their membership functions for: Goal distance

Table 1.4 Horizontal navigation control rules

Goal distance	ZERO	NEAR	MID	FAR
Horizontal speed				
ZERO	ZERO	ZERO	PMAX	PMAX
SLOW	NMIN	PMIN	PMAX	PMAX
FAST	NMID	PMIN	PMID	PMAX
WARP	NMAX	NMIN	PMIN	PMAX

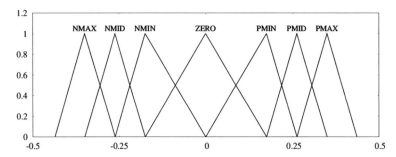

Fig. 1.30 Output linguistic variables and their membership functions for: OPitchNavigation (the Pitch setpoint)

direction, making a hexacopter to fly forwards. It is important to note that, by using the mentioned inputs, horizontal navigation fuzzy controller commands the horizontal movement in a smooth way, i.e. as the hexacopter comes close to the target point, its horizontal speed decreases in order to alleviate the control bounce produced by the movement inertia.

The horizontal navigation control surface is depicted in Fig. 1.31. When the hexacopter is on the target position, the goal distance is near to zero and the horizontal speed is zero, the pitch setpoint is zero. In this state the pitch controller stabilizes the hexacopter on the current position, and hence, hexacopter is kept with pitch angle at zero. When the goal distance is greater than zero, this controller changes the pitch setpoint.

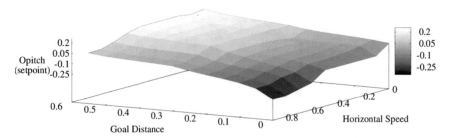

Fig. 1.31 Fuzzy control surface for horizontal navigation control

Thus, the pitch controller stabilizes the hexacopter in new pitch angle, which higher than zero, making the hexacopter to move forward.

When the goal distance is zero and the horizontal speed is high, the output is negative. In this situation the vehicle is moving while passing over the target position, and hence, it must slow down. This is the reason why this controller output must be negative.

1.4.6 The Vertical Navigation

Vertical navigation fuzzy controller is similar to the horizontal navigation controller. However, it controls the movement on the Z-axis. Figure 1.32 shows the block diagram of this controller. This controller takes as input the vertical distance to the target point, as well as the vertical speed.

The first one is the error in the Euler approximation of Z-axis, while the second one is the difference in distance (Z-axis). Linguistic variables for vertical distance and vertical speed are presented in Figs. 1.33 and 1.34, respectively.

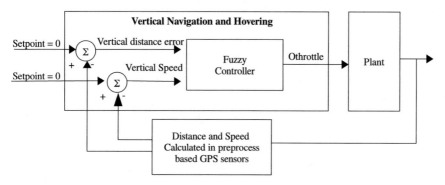

Fig. 1.32 The vertical navigation fuzzy controller

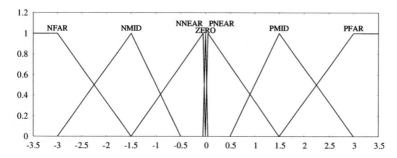

Fig. 1.33 Input linguistic variables and their membership functions for: Vertical distance

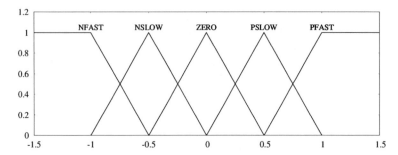

Fig. 1.34 Input linguistic variables and their membership functions for: Vertical speed

Table 1.5 Vertical navigation control rules

V. dist.	NFAR	NMID	NNEAR	ZERO	PNEAR	PMID	PFAR
V. speed							
NFAST	NMAX	NMID	NMIN	PMIN	PMID	PMAX	PMAX
NSLOW	NMAX	NMID	NMIN	PMIN	PMID	PMAX	PMAX
ZERO	NMAX	NMID	NMIN	ZERO	PMIN	PMID	PMAX
PSLOW	NMAX	NMAX	NMID	NMIN	PMIN	PMID	PMAX
PFAST	NMAX	NMAX	NMID	NMIN	PMIN	PMID	PMAX

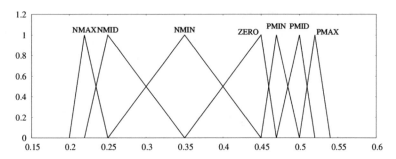

Fig. 1.35 Output linguistic variables and their membership functions for: OThrottle

Table 1.5 shows the 35 rules that compose the vertical navigation fuzzy controller. As the result, this controller sets the omega throttle variable (OThrottle), presented in Fig. 1.35 which is decomposed in the amount of power applied on all rotors, increasing or decreasing the overall lift force making the hexacopter fly on higher or lower altitude. It is worth noting that this presents a smooth control approach similar to the horizontal navigation, i.e. the power applied on the rotors decreases along with vertical speed as the hexacopter comes closer to target altitude.

The fuzzy surface control for vertical navigation and hovering is shown in Fig. 1.36. The altitude is maintained by controlling the throttle applied onto the all rotors. The input information is taken from the GPS sensor. The vertical speed is

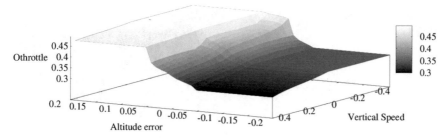

Fig. 1.36 Fuzzy control surface for vertical navigation and hovering control

used to avoid the hexacopter to oscillate up and down. It is similar as the acceleration is used to avoid oscillation in the pitch and roll controller. It is worth to note that the ZERO output not means zero value, but the value that the hexacopter is hovering. To realize the relationship between altitude error and vertical speed, suppose the altitude error is zero, the hexacopter is in the target vertical position, but also suppose the vertical speed is positive, perhaps 0.4 or higher. It means the hexacopter reached the target and goes beyond because it is in movement to up. It must be slowed down. Therefore, the controller sets the output value to a value lower than the ZERO, causing the hexacopter to slow down. On the other hand, when the altitude is zero and the vertical speed is negative it means the hexacopter is falling down. In this condition, the controller must to set output to a value higher than the ZERO just to make the hexacopter stop the falling.

1.5 Experiment and Results

1.5.1 Overview of the Experiment

The work has been validated through a case study by means of simulation. For that, a model of a hexacopter has been created in the V-REP robotics simulation environment. A free payload has been attached to the hexacopter forming a pendulum as depicted in Fig. 1.37. In the simulated environment, the hexacopter weighs 980 g (mass = 0.1) and the payload weighs 49 g (mass = 0.005). The hexacopter model used is one that is already available on V-REP. Such a payload weight was defined in 5 % of the hexacopter weight due to limitations on the rotors model that cannot provide enough thrust to allow the hexacopter takeoff. For simulation the V-REP has been configured with "Dynamic engine" as "Bullet", the "Dynamics settings" as "Verry accurate" and "Simulation time step" as "dt = 10 ms".

The major challenge tackled in this work is the switching among fuzzy controllers at the right moment. For instance, let us assume that a hexacopter starts on the ground and receives a command to fly to a certain position, e.g. 3 m in latitude, 3 m

Fig. 1.37 Moving carrying a
weight

in longitude and 3 m upward. To achieve the commanded position, every movement
must be executed properly, and hence, all controllers must work cooperatively to
achieve the goal. In other words, the fuzzy controller that controls the longitudinal
position cannot override the other controllers actions. On the other hand, when the
fuzzy controllers do not work well together, a controller output could take the rotors
actuation over the output of the other controller **controllers' outputs might overlap
each other**. In this case, there must exist a high-level controller that performs the
contention of misbehaved controllers.

The proposed approach has been validated through simulation. For that, a model of
a hexacopter has been created in the V-REP simulation environment. A free payload
has been attached to the hexacopter forming a pendulum. Thus, the proposed multi-
layer controller must control the hexacopter movement when it is commanded to
move to another position, while keeping the its body stabilized during the flight. In
other words, the results indicate that the proposed approach is robust since it allows
the hexacopter move from one position to another, even though it must carry a moving
payload.

The proposed multi-layer fuzzy controller has been implemented on Linux using
the C language and gcc compiler version 4.9.1. This software communicates with
V-REP environment, receiving the input sensor signals and sending the output commands generated by the proposed fuzzy controller. The main idea was to develop
hybrid fuzzy controllers. The control system is composed of three threads: (i) a thread
for the main control loop; (ii) a thread to produce a data log which was used to create
the charts presented in this section; and (iii) another one to insert the user commands
as target point coordinates used as setpoints to the the proposed controller. Therefore,
the developed software acts as both the hexacopter movement and stability controller
and the interface between the operator and the hexacopter.

The experiment has been performed in two phases. The first one has concentrated
on calibrating the range of values for all linguistic variables of the five fuzzy controllers. The results of this phase was described in see Sect. 1.4. Those values have

Table 1.6 The position sent to hexacopter as command

Commands	X	Y	Z
1st	+3	+3	+3
2nd	−2	+1	+5
3rd	+4	−2	+8
4th	−2	−1	+2
5th	0	0	+2
6th	0	0	0

been defined in a manual and arbitrary way by means of an iterative trial and error process. In this phase, the aim was to achieve a stabilized movement and hovering for the hexacopter. For that, the hexacopter has flown without attached payload. The hexacopter has been commanded to fly to six distinct positions from the origin point (i.e. X = 0, Y = 0, Z = 0) by means of the following relative coordinates expressed in meters shown in Table 1.6.

The ranges for each value of all linguistic variables have been defined using an intuitive try-and-error method. On each simulation round, threshold limits have been tuned until the hexacopter was able to fly and hover on a fixed point in a stable way.

The second phase focused on evaluating how the proposed multi-level fuzzy controller behaves when the hexacopter carries a free payload. In other words, this phase assesses how the proposed controller behaves in situations of stress caused by the pendulum effect created by the inertial movement of the free payload. The same commands have been issue in the same order as described earlier. All data generated during the simulation have been analyzed and the results are discussed in the next section.

1.5.2   Results

This section presents the results in terms of how the values of the controlled variables evolved during the experiment. As mentioned, once the hexacopter receives the command to fly to a new position, it performs the following sequence of actions: (1) the hexacopter flies up until reaching the target altitude; (2) the controller established the new yaw angle in order to aim at the desired X- and Y-axis position; and, finally, (3) the hexacopter flies horizontally toward the target position. This sequence of steps is obtained by establishing thresholds in the transitions of motion events, as depicted in Fig. 1.11. Figure 1.38 shows the footprint of the trajectory flown by the hexacopter during the experiment.

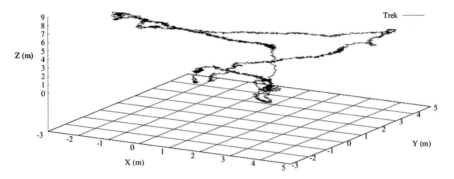

Fig. 1.38 Flight footprint

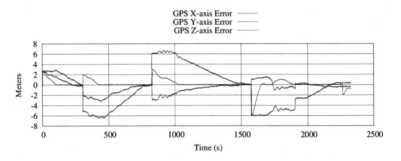

Fig. 1.39 Movement and hovering stabilization control

Figures 1.39 shows the error of the GPS coordinates that have been read during the entire flight. The moment when a new command has been received by the hexacopter, please refer to Table 1.6: (i) the 1st command was received at 0 s; (ii) the 2nd command at 300 s; (iii) the 3rd command at 800 s; (iv) the 4th command at 1600 s; (v) the 5th command at 2500 s;

Videos of the experiment can be seen on:

https://www.youtube.com/watch?v=dTJhH8lU6BY.

Flying data generated during the simulation can be observed on Figs. 1.40, 1.41, 1.42 and 1.43.

1.5.3 Discussion

It is worth noting that, despite some disturbance caused by the definition of a new position, the hexacopter moved smoothly and in a stable way from the current position to the target position. In addition, the proposed multi-layer fuzzy controller has responded properly to the control stress imposed by the free payload. Such a claim is

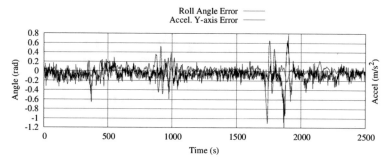

Fig. 1.40 Roll stabilization—inputs

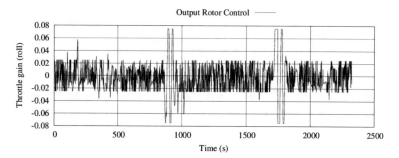

Fig. 1.41 Roll stabilization—output

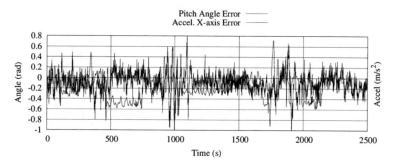

Fig. 1.42 Pitch stabilization—input

supported by the analysis of flying data plotted on the charts presented in Figs. 1.40, 1.41, 1.42, and 1.43.

In these charts, at moments close to 300, 800, 1600 and 2500 s, the present variation seems to indicate a poor stability.

However, at each time instant, a new target position is sent to the hexacopter, disturbing the system. Therefore, the controllers must act to control the hexacopter stability. The output variable value of these controllers varies around $+/-0.08$ representing 8 % variation within 150 s, which is considered acceptable for a stable

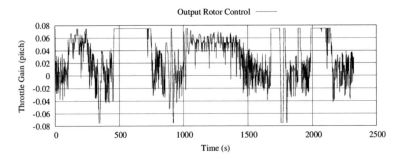

Fig. 1.43 Pitch stabilization—output

flight of a hexacopter carrying a free payload. A video of the simulation shows the hexacopter flying stable, as well as it shows the hexacopter recovering from these disturbances.

Furthermore, tuning membership functions took a considerable time. However, once they are correctly calibrated with the unloaded hexacopter, the proposed fuzzy controller was able to control the movement and the stability without modifications. Thus, it was observed the flexibility of fuzzy logic for designing a complex control systems as the one presented in this work. Other perceived advantage of fuzzy logic is the handling of non-linear scale independently from input or output, for instance the scale for the throttle output, goal distance, pitch navigation output.

During the calibration phase, various models have been tried in order to properly control the hexacopter stability during horizontal and vertical navigation – the stabilized hovering has been achieved easily. It was not enough to sum the value OPitch and ORoll to the throttle variable (OThrottle). It has been observed that V-REP simulates random errors while the simulation is running as it happens in the real world. Such errors affect the throttle proportionally, and hence, to keep the roll and pitch stabilization the outputs of these fuzzy controllers must follow this growth. To cope with this situation, outputs of pitch and roll controls were modified to represent a gain based on a percentage of the current throttle (see Eqs. 1.7–1.12).

1.6 Conclusions and Future Work

This work proposes an approach that integrates several fuzzy controller that work collaboratively to keep the stabilization and control the navigation of an Hexacopter carrying a free payload while it is flying and hovering. This paper discusses how the propose controller has been designed.

In order do evaluate the proposed approach, a simulation experiment has been conducted. The proposed approach was able to stabilize the hovering, control the altitude, position, and navigation of the simulated hexacopter. Additionally, the proposed multi-layer fuzzy controller has responded properly to the control stress imposed by the free payload. When the hexacopter has flown with the free payload, the control system took more time to stabilize. By using fuzzy logic, it was possible to see the flexibility of the proposed approach, since once the variables have been calibrated, it was not necessary to change the systems to allow the hexacopter to fly stable carrying an attached free payload. However, on the other hand, the side-effect of this flexibility is the difficult to tune the variables thresholds.

The next steps involve the implementation of the proposed multi-layer fuzzy controller on a real Hexacopter. A computing systems with several distinct sensors is envied so that it allows moving towards a fully autonomous UAV that executes localization, map building, path planning and mission execution. In addition, considering the difficult of tuning the proposed systems in the calibration phase, other future work direction is to develop adaptive fuzzy control to assist in this task.

References

1. Ahmed OA, Latief M, Ali MA, Akmeliawati R (2015) Stabilization and control of autonomous hexacopter via visual-servoing and cascaded-proportional and derivative (PD) controllers. 2015 6th international conference on automation, robotics and applications (ICARA), pp 542–549. IEEE
2. Alaimo A, Artale V, Milazzo CLR, Ricciardello A (2014) PID controller applied to hexacopter flight. J Intell Robot Syst 73:261–270
3. Bacik J, Perdukova D, Fedor P (2015) Design of fuzzy controller for hexacopter position control. artificial intelligence perspectives and applications. Advances in intelligent systems and computing. Springer International Publishing, Berlin, pp 192–202
4. Bipin K, Duggal V, Madhava K (2015) Autonomous navigation of generic monocular quadcopter in natural environment. Autonomous navigation of generic monocular quadcopter in natural environment, pp 1063–1070. IEEE
5. Collotta M, Pau G, Caponetto R (2014) A real-time system based on a neural network model to control hexacopter trajectories. 2014 international symposium on power electronics, electrical drives, automation and motion (SPEEDAM)
6. Genetic algorithm applied to the stabilization control of a hexarotor (2015) Genetic algorithm applied to the stabilization control of a hexarotor. In: Proceedings of the international conference on numerical analysis and applied mathematics, vol 1648
7. Haque Md R, Muhammad M, Swarnaker D, Arifuzzaman M (2014) Autonomous Quadcopter for product home delivery. 2014 international conference on electrical engineering and information & communication technology (ICEEICT), pp 1–5. IEEE
8. Lee EA, Seshia SA (2015) Introduction to embedded systems – a cyber-physical systems approach, 2nd edn. Springer, New York www.leeSeshia.org, Chapters 1–3
9. Leishman R, Macdonald J, McLain T, Beard R (2012) Relative navigation and control of a hexacopter. Autonomous Quadcopter for product home delivery. 2012 IEEE international conference on robotics and automation (ICRA)
10. Ma sum MA, Jati G, Arrofi MK, Wibowo A, Mursanto P, Jatmiko W (2013) Autonomous quadcopter swarm robots for object localization and tracking. 2013 international symposium on micro-nanomechatronics and human science (MHS)
11. Ołdziej D, Gosiewski Z (2013) Modelling of Dynamic and control of six-rotor autonomous unmanned aerial vehicle. solid state phenomena. Trans Tech Publ, pp 220–225
12. Passino KM, Yurkvich S (1998) Fuzzy control, 1.2–conventional control system design. Addison-wesley, Boston Subsection 2.2

13. Rohmer E, Singh SPN, Freese M (2013) V-REP: A versatile and scalable robot simulation framework. 2013 IEEE/RSJ international conference on intelligent robots and systems (IROS), pp 1321–1326. IEEE
14. Salih AL, Moghavvemi M, Mohamed HAF, Gaeid KS (2010) Scientific Research and Essays. Flight PID controller design for a UAV quadrotor. Academic Journals, Denmark
15. Siegwart R, Nourbakhsh IR, Scaramuzza D (2011) Introduction to autonomous mobile robots. MIT press, Cambridge

Chapter 2
Monocular Pose Estimation for an Unmanned Aerial Vehicle Using Spectral Features

Gastón Araguás, Claudio Paz, Gonzalo Perez Paina and Luis Canali

Pose estimation of Unmanned Aerial Vehicles (UAV) using cameras is currently a very active research topic in computer and robotic vision, with special application in GPS-denied environments. However, the use of visual information for ego-motion estimation presents several difficulties, such as features search, data association (feature correlation), inhomogeneous features distribution in the image, etc. We propose a visual position and orientation estimation algorithm based on the discrete homography constraint, induced by the presence of planar scenes, and the so-called spectral features in the image. Our approach has the following unique characteristics: it selects the appropriate distribution of the features in the image, it does not need either initialization process or search for features, and it does not depend on the presence of corner-like features in the scene. The position and orientation estimation is made using a down-looking monocular camera rigidly attached to a quadrotor. It is assumed that the floors over which the quadrotor flights are planar, and therefore two consecutive images are related by a homography induced by the floor plane. This homography constraint is more appropriate than the well-known epipolar constraint, which vanishes for a zero translation and loses rank in the case of planar scenes. The

G. Araguás (✉) · C. Paz · G.P. Paina · L. Canali
Centro de Investigación en Informática para la Ingeniería (CIII), Facultad
Regional Córdoba, Universidad Tecnológica Nacional, Córdoba, Argentina
e-mail: garaguas@frc.utn.edu.ar

C. Paz
e-mail: cpaz@frc.utn.edu.ar

G.P. Paina
e-mail: gperez@frc.utn.edu.ar

L. Canali
e-mail: lcanali@frc.utn.edu.ar

© Springer International Publishing Switzerland 2017
N. Nedjah et al. (eds.), *Designing with Computational Intelligence*,
Studies in Computational Intelligence 664,
DOI 10.1007/978-3-319-44735-3_2

pose estimation algorithm is tested in a simulated dataset, and the robustness of the spectral features is evaluated in different conditions using a conveyor belt.

2.1 Introduction

In the last years quadrotors have gained popularity in entertainment, aero-shooting and many other civilian or military applications, mainly due to their low cost and great controllability. Between other tasks, they are a good choice for operation at low altitude, in cluttered scenarios or even for indoor applications. Such environments limit the use of GPS or compass measurements which are indeed excellent options for attitude determination in wide open outdoor areas [1, 12]. These constraints have motivated, over the last years, the extensive use of on-board cameras as a main sensor for state estimation [5, 14, 16]. In this context, we present a new approach to estimate the ego-motion of a quadrotor in indoor environments for smooth flights, using a down-looking camera for translation and rotation calculation. As a continuation of the work presented in [3], we propose the utilization of a fixed number of patches distributed on each image of the sequence to determine the ego-motion of the camera, based on the plane-induced homography that relates the patches in two consecutive frames.

A number of spatial and frequency domain approaches have been proposed to estimate the image-to-image transformation, between two views of a planar scene, most of them limited to similarities. Spatial domain methods need corresponding points, lines, conics, etc. [7, 9, 10], whose identification in many practical situations is non-trivial, thereby limiting their applicability. Scale, rotation, and translation invariant features have been popular facilitating recognition under these transformations. Geometry of multiple views of the same scene has been a subject of extensive research over the past decade. Important results relating corresponding entities such as points and lines can be found in [7, 9]. Recent work has also focused on more complex entities such as conics and higher-order algebraic curves [10]. However, these approaches depend on extracting corresponding entities such as points, lines or contours and do not use the abundant information present in the form of the intensity values in the multiple views of the scene. Frequency domain methods are in general superior to methods based on spatial features because the entire image information is used for matching. They also avoid the crucial issue regarding the selection of the best features.

Our work proposes the use of a fixed number of patches distributed on each image of the sequence to determine the pose change of a moving camera. The pose of the camera (and UAV) is estimated trough dead-reckoning, performing a time integration of ego-motion parameters determined between frames. We concentrate in the XY-position and the orientation estimation in order to fuse these parameters with the on-board IMU and altimeter sensors measurements. The camera ego-motion is estimated using the homography induced by the (assumed to be flat) floor, and the corresponding points needed to estimate the homography are obtained on the frequency domain. A point in the image is represented by the spectral information of an image patch, which we call spectral feature [2, 3]. The correspondence between points in two consecutive frames is determined by means of the phase correlation between each spectral feature pair. These kind of features perform better than the interest points based on the image intensity when observing a floor with homogeneous texture. Moreover, since their position in the image plane is previously selected, they are always well distributed.

The transformation that relates two images taken from different views (with a moving camera) contains information about the spatial rotation and translation of the views, or the camera movement. Considering a downward-looking camera, and assuming that the floor is a planar surface, all the space points imaged by the camera are coplanar and there is a homography between the world and the image planes. Under this constraint, if the camera center moves, the images taken from different points of view are also related by a homography. The spatial transformation that relates both views can be completely determined from this homography between images.

The chapter is organized as follows: Sect. 2.2 details the homography-based pose estimation, with a review of the so-called plane-induced homography. In this section the homography decomposition used to obtain the translation and rotation of the camera is also presented; and in order to estimate the homography, the so-called spectral features are introduced in Sect. 2.3. The implementation details and the results are presented in Sect. 2.4, and finally Sect. 2.5 remarks the conclusions and future work.

2.2 Homography-Based Pose Estimation

The visual pose estimation is based on the principle that two consecutive images of a planar scene are related by a homography. The planar scene corresponds to the floor surface, which is assumed to be relatively flat, observed by the down-looking camera on the UAV. The spatial transformation of the camera, and therefore of the UAV, is encoded in this homography. Knowing the homography matrix that relates both images, the transformation parameters that describe the camera rotation and translation can be obtained.

In order to estimate the homography induced by the planar surface, a set of corresponding points on two consecutive images must be obtained. This process is performed selecting a set of features in the first image and finding the corresponding set of features in the second one. Then, the image coordinates of each feature in both images conform the set of corresponding image points needed to calculate the homography.

The image features used in our approach are the so-called spectral features, a Fourier domain representation of an image patch. Selecting a set of patches in both images (the same number, with the same size and position), the displacement between them is proportional to the phase shift between the associated spectral features, and

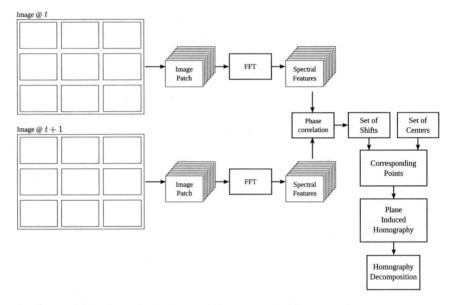

Image @ t

Image @ $t+1$

Fig. 2.1 Block diagram of the implemented visual pose estimation approach

can be obtained using the Fourier shift theorem. This displacement, in addition to the feature center, determines the correspondence between features in both images: that is, the set of corresponding points needed to estimate the homography.

In Fig. 2.1 a block diagram of the estimation process is shown. Here, as an example, nine spectral features in both images are used.

2.2.1 Review of Plane-Induced Homography

Given a 3D scene point **P**, and two coordinate systems, CS_A and CS_B, the coordinates of the point **P** on each one can be denoted by $\mathbf{X}_A$ and $\mathbf{X}_B$ respectively. If $R_A^B \in SO(3)$ is the rotation matrix that changes the representation of a point in CS_A to CS_B, and $\mathbf{T}_B \in \mathbb{R}^{3\times1}$ is the translation vector of the origin of CS_A w.r.t CS_B (expressed in CS_B), then the representations of the point **P** relate each other as

$$\mathbf{X}_B = R_A^B \mathbf{X}_A + \mathbf{T}_B. \tag{2.1}$$

We suppose now that the point **P** belongs to a plane π, denoted in the coordinate system CS_A by its normal $\mathbf{n}_A$ and its distance to the coordinate origin d_A. Therefore, the following plane equation holds

$$(\mathbf{n}_A)^T \mathbf{X}_A = d_A \qquad \Rightarrow \qquad \frac{(\mathbf{n}_A)^T \mathbf{X}_A}{d_A} = 1, \tag{2.2}$$

Plugging (2.2) into (2.1) we have

$$\mathbf{X}_B = \left(R_A^B + \frac{\mathbf{T}_B}{d_A} (\mathbf{n}_A)^T \right) \mathbf{X}_A = H_A^B \mathbf{X}_A, \tag{2.3}$$

with

$$H_A^B \doteq \left(R_A^B + \frac{\mathbf{T}_B}{d_A} (\mathbf{n}_A)^T \right). \tag{2.4}$$

The matrix H_A^B is a *plane-induced homography*, in this case induced by the plane π. As can be seen, this matrix encodes the transformation parameters that relate both coordinates systems (R_A^B and $\mathbf{T}_B$), and the structure parameters of the environment ($\mathbf{n}_A$ and d_A).

Considering now a moving camera associated to the coordinate system CS_A at time t_A and by CS_B at time t_B, according to the central projection model the relations between the 3D points and their projections on the camera normalized plane are given by

$$\lambda_A \mathbf{x}_A = \mathbf{X}_A; \qquad\qquad \lambda_B \mathbf{x}_B = \mathbf{X}_B \tag{2.5}$$

where $\lambda_A \in \mathbb{R}^+$ and $\lambda_B \in \mathbb{R}^+$. Using (2.5) in Eq. (2.3) we have

$$\lambda_B \mathbf{x}_B = H_A^B \lambda_A \mathbf{x}_A \qquad \Rightarrow \qquad \mathbf{x}_B = \lambda H_A^B \mathbf{x}_A, \tag{2.6}$$

with $\lambda = \frac{\lambda_A}{\lambda_B}$. Given that both vectors $\mathbf{x}_B$ and $\lambda H_A^B \mathbf{x}_A$ have the same direction

$$\mathbf{x}_B \times \lambda H_A^B \mathbf{x}_A = \hat{\mathbf{x}}_B H_A^B \mathbf{x}_A = 0, \tag{2.7}$$

with $\hat{\mathbf{x}}_B$ the skew-symmetric matrix associated to $\mathbf{x}_B$. The Eq. (2.7) is known as the *planar epipolar restriction*, and holds for all 3D points belonging to the plane π. Assuming that the camera is pointing to the ground (downward-looking camera) and that the scene structure is approximately a planar surface, all the 3D points captured by the camera will fulfill this restriction.

The homography H_A^B represents the transformation of the camera coordinate systems between instant t_A and t_B, hence, it contains the information of the camera rotation and translation between these two instants. This homography can be estimated knowing at least four corresponding points of two images. In our case the correspondence between these points is calculated in the spectral domain, by means of the *spectral features*. The complete process is detailed in Sect. 2.3.

2.2.2 Homography Decomposition

Following [13] H can be decomposed in order to obtain a non-unique solution (exactly four different solutions) $\left\{R_i, \mathbf{n}_i, \frac{\mathbf{T}_i}{d_i}\right\}$. Then, adding some extra data for disambiguation we can arrive to the appropriate $\left\{R_A^B, \mathbf{n}_A, \frac{\mathbf{T}_B}{d_A}\right\}$ solution.

Normalization

Given that the planar epipolar constraint ensures equality only in the direction of both vectors (Eq. (2.7)), what is actually obtained after the homography estimation is λH, that is[1]

$$H_\lambda = \lambda H = \lambda \left(R + \frac{\mathbf{T}}{d}\mathbf{n}^T\right). \tag{2.8}$$

The unknown factor λ included in H_λ can be found as follows. Consider the product

$$H_\lambda^T H_\lambda = \lambda^2 (I + Q) \tag{2.9}$$

with I the identity, $Q = \mathbf{a}\mathbf{n}^T + \mathbf{n}\mathbf{a}^T + ||\mathbf{a}||^2\mathbf{n}\mathbf{n}^T$ and $\mathbf{a} = \frac{1}{d}R^T\mathbf{T} \in \mathbb{R}^{3\times1}$. The vector $\mathbf{a} \times \mathbf{n}$, perpendicular to $\mathbf{a}$ and $\mathbf{n}$, is an eigenvector of $H_\lambda^T H_\lambda$ associated to the eigenvalue λ^2, being that

$$H_\lambda^T H_\lambda(\mathbf{a} \times \mathbf{n}) = \lambda^2(\mathbf{a} \times \mathbf{n}). \tag{2.10}$$

So, if λ^2 is an eigenvalue of $H_\lambda^T H_\lambda$, then $|\lambda|$ is a singular value of H_λ. It is easy to show that Q in (2.9) has one positive, one zero and one negative eigenvalue, what means that λ^2 is the second ordereigenvalue of $H_\lambda^T H_\lambda$, and $|\lambda|$ will be the second order singular value of H_λ. That is, if $\sigma_1 > \sigma_2 > \sigma_3$ are the singular values of H_λ, then

$$H = \pm\frac{H_\lambda}{\sigma_2} \tag{2.11}$$

To get the right sign of H, the positive depth condition in (2.6) must be applied. In order to ensure that all the considered points are in front of the camera, all 3D points in plane π projected in the image plane must fulfill

$$(\mathbf{x}_B^j)^T H\mathbf{x}_A^j = \frac{1}{\lambda_j} > 0, \quad \forall j = 1, 2, \ldots, n. \tag{2.12}$$

where $\left(\mathbf{x}_A^j, \mathbf{x}_B^j\right)$ are the projections of all points $\{\mathbf{P}\}_{j=1}^n$ lying on the plane π, at time t_A and t_B respectively.

---

[1]To avoid the abuse of notation we do not use here the sub and supra indexes A and B that refer to the corresponding coordinate systems.

Estimation of n

The homography H induced by the plane π preserves the norm of any vector in the plane, i.e. given a vector $\mathbf{r}$ such that $\mathbf{n}^T\mathbf{r} = 0$, then

$$Hr = Rr \tag{2.13}$$

and therefore $||H\mathbf{r}|| = ||\mathbf{r}||$. Consequently, knowing the space spanned by the vectors that preserve the norm under H, the perpendicular vector $\mathbf{n}$ is also known.

The matrix $H^T H$ is symmetric, and therefore admits eigenvalue decomposition. Being $\sigma_1^2, \sigma_2^2, \sigma_3^2$ the eigenvalues and $\mathbf{v}_1, \mathbf{v}_2, \mathbf{v}_3$ the eigenvectors of $H^T H$, then

$$H^T H\mathbf{v}_1 = \sigma_1^2 \mathbf{v}_1, \quad H^T H\mathbf{v}_2 = \mathbf{v}_2,$$
$$H^T H\mathbf{v}_3 = \sigma_3^2 \mathbf{v}_3 \tag{2.14}$$

since by the normalization $\sigma_2^2 = 1$. That is, $\mathbf{v}_2$ is perpendicular to $\mathbf{n}$ and $\mathbf{T}$, so its norm is preserved under H. From (2.14) it can be shown that the norm of the following vectors

$$\mathbf{u}_1 \doteq \frac{\sqrt{1-\sigma_3^2}\mathbf{v}_1 + \sqrt{\sigma_1^2-1}\mathbf{v}_3}{\sqrt{\sigma_1^2-\sigma_3^2}},$$
$$\mathbf{u}_2 \doteq \frac{\sqrt{1-\sigma_3^2}\mathbf{v}_1 - \sqrt{\sigma_1^2-1}\mathbf{v}_3}{\sqrt{\sigma_1^2-\sigma_3^2}} \tag{2.15}$$

is preserved under H too, as well as all vectors in the sub-spaces spanned by

$$S_1 = \text{span}\{\mathbf{v}_2, \mathbf{u}_1\}, \quad S_2 = \text{span}\{\mathbf{v}_2, \mathbf{u}_2\} \tag{2.16}$$

Therefore, there exist two possible planes that can induce the homography H, π_1 and π_2, defined by the normal vectors to S_1 and S_2

$$\mathbf{n}_1 = \mathbf{v}_2 \times \mathbf{u}_1, \quad \mathbf{n}_2 = \mathbf{v}_2 \times \mathbf{u}_2. \tag{2.17}$$

Estimation of R

The action of H over $\mathbf{v}_2$ and $\mathbf{u}_1$ is equivalent to a pure rotation

$$H\mathbf{v}_2 = R_1\mathbf{v}_2, \quad H\mathbf{u}_1 = R_1\mathbf{u}_1 \tag{2.18}$$

since both vectors are orthogonal to $\mathbf{n}_1$. The rotation of $\mathbf{n}_1$ can be computed as

$$R_1\mathbf{n}_1 = H\mathbf{v}_2 \times H\mathbf{u}_1. \tag{2.19}$$

Defining the matrix $U_1 = [\mathbf{v}_2, \mathbf{u}_1, \mathbf{n}_1]$ and $W_1 = [H\mathbf{v}_2, H\mathbf{u}_1, H\mathbf{v}_2 \times H\mathbf{u}_1]$, from (2.18) and (2.19) we have

$$R_1 U_1 = W_1 \tag{2.20}$$

and given that the set of vectors $\{\mathbf{v}_2, \mathbf{u}_1, \mathbf{n}_1\}$ form an orthogonal base in $\mathbb{R}^3$, the matrix U_1 is non-singular, therefore

$$R_1 = W_1 U_1^T, \tag{2.21}$$

that is

$$R_1 = [H\mathbf{v}_2, H\mathbf{u}_1, H\mathbf{v}_2 \times H\mathbf{u}_1][\mathbf{v}_2, \mathbf{u}_1, \mathbf{n}_1]^T. \tag{2.22}$$

Considering now the set $\{\mathbf{v}_2, \mathbf{u}_2, \mathbf{n}_2\}$, in the same way we arrive to

$$R_2 = W_2 U_2^T \tag{2.23}$$

where $U_2 = [\mathbf{v}_2, \mathbf{u}_2, \mathbf{n}_2]$ and $W_2 = [H\mathbf{v}_2, H\mathbf{u}_2, H\mathbf{v}_2 \times H\mathbf{u}_2]$, that is

$$R_2 = [H\mathbf{v}_2, H\mathbf{u}_2, H\mathbf{v}_2 \times H\mathbf{u}_2][\mathbf{v}_2, \mathbf{u}_2, \mathbf{n}_2]^T. \tag{2.24}$$

Estimation of $\frac{\mathbf{T}}{d}$

Once R and $\mathbf{n}$ are known, the estimation of $\frac{\mathbf{T}}{d}$ is direct, as

$$\frac{\mathbf{T}_1}{d_1} = (H - R_1)\,\mathbf{n}_1, \tag{2.25}$$

$$\frac{\mathbf{T}_2}{d_2} = (H - R_2)\,\mathbf{n}_2, \tag{2.26}$$

which completes both solutions of the H decomposition.

Desambiguation

However, it should be noted that the term $\frac{\mathbf{T}}{d}\mathbf{n}^T$ in H introduces a sign ambiguity, since $\frac{\mathbf{T}}{d}\mathbf{n}^T = \frac{-\mathbf{T}}{d}(-\mathbf{n}^T)$, therefore the number of possible solutions rises to four,

$$\left\{R_1, \mathbf{n}_1, \frac{\mathbf{T}_1}{d_1}\right\}, \quad \left\{R_1, -\mathbf{n}_1, \frac{-\mathbf{T}_1}{d_1}\right\}, \\ \left\{R_2, \mathbf{n}_2, \frac{\mathbf{T}_2}{d_2}\right\}, \quad \left\{R_2, -\mathbf{n}_2, \frac{-\mathbf{T}_2}{d_2}\right\}. \tag{2.27}$$

In order to ensure that the plane inducing the homography H appears in front of the camera, each normal vector $\mathbf{n}_i$ must fulfill $n_z < 0$, and therefore only two solutions remain. These two solutions are both physically possible, but given that most of the time the camera on the UAV is facing-down, we choose the solution with the normal vector $\mathbf{n}$ closest to $[0, 0, -1]^T$ in terms of the norm L_2.

2.3 Spectral Features Correspondence

The estimation of the homography given by two consecutive images from a moving camera requires a set of corresponding points. Classically, this set of points is obtained by detecting features, such as lines and corners in both images, and determining correspondences. The feature detectors are typically based on image gradient methods. An alternative to this approach is to use frequency-based features, or spectral features, and to determine correspondences in the frequency domain.

The so-called spectral feature refers to the Fourier domain representation of an image patch of $2^n \times 2^n$, where $n \in \mathbb{N}^+$ is set accordingly to the allowed image displacement [3]. The power of 2 of this patch size is selected based on the efficiency of the Fast Fourier Transform (FFT) algorithm. The number and position of spectral features in the image are set beforehand. Even though a minimum of four points are needed to estimate the homography, a higher number of features are used to increase the accuracy, and the RANSAC algorithm [8] is used for outliers elimination.

Consider two consecutive frames, where spectral features on each image were computed. To determine the correspondence between features is equivalent to determine the displacement between them. This displacement can be obtained using the spectral information by means of the Phase Correlation Method (PCM) [11]. This method is based on the Fourier shift theorem, which states that the Fourier transforms of two identical but displaced images differ only in a phase shift.

Given two images i_A and i_B of size $N \times M$ differing only in a displacement (u, v), such as

$$i_A(x, y) = i_B(x - u, y - v) \tag{2.28}$$

where

$$u \le x < N - u, \, v \le y < M - v, \tag{2.29}$$

their Fourier transforms are related by

$$I_A(\omega_x, \omega_y) = e^{-j(u\omega_x + v\omega_y)} I_B(\omega_x, \omega_y), \tag{2.30}$$

where I_A and I_B are the Fourier transforms of images i_A and i_B, respectively; u and v are the displacements for each axis. From (2.30), the amplitudes of both transformations are the same and only differ in phase which is directly related to the image displacement (u, v), and therefore this displacement can be obtained using the cross-power spectrum (CPS) of the given transformations I_A and I_B. The CPS of two complex functions is defined as

$$\mathcal{C}(F, G) = \frac{F(\omega_x, \omega_y) G^*(\omega_x, \omega_y)}{|F(\omega_x, \omega_y)||G^*(\omega_x, \omega_y)|} \tag{2.31}$$

where G^* is the complex conjugate of G.

Using (2.30) in (2.31) over the transformed images I_A and I_B, gives

$$\frac{I_A I_B^*}{|I_A||I_B^*|} = e^{-j(u\omega_x + v\omega_y)}. \qquad (2.32)$$

The inverse Fourier transform of (2.32) is an impulse located exactly in (u, v), which represents the displacement between the two images

$$\mathcal{F}^{-1}[e^{-j(u\omega_x + v\omega_y)}] = \delta(x - u, y - v). \qquad (2.33)$$

Using the discrete Fast Fourier Transform (FFT) algorithm instead of the continuous version, the result will be a pulse signal centered in (u, v) [17].

2.3.1   Corresponding Points

The previous subsection describes how to calculate the displacement between two images using PCM. Applying this method to each image patch pair, the displacement between spectral features is determined. The set of corresponding points required to estimate the homography can be constructed with the patch centers of the first image and the displaced patch centers of the second one, that is

$$\{\mathbf{x}_{A_i} \leftrightarrow \mathbf{x}_{A_i} + \Delta\mathbf{d}_i = \mathbf{x}_{B_i}\} \qquad (2.34)$$

where $\Delta\mathbf{d}_i$ represents the displacement between the i-th spectral feature, and $\mathbf{x}_{A_i}$ the center of the i-th spectral feature in the CS_A. This is schematically shown in the zoomed area of Fig. 2.2. As shown in Sect. 2.2.1, this set of corresponding points is

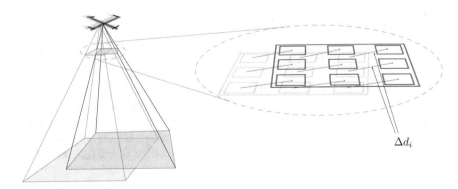

Fig. 2.2 Estimation of the rotation and translation between two consecutive images based on spectral features

Fig. 2.3 Displacements
between patches

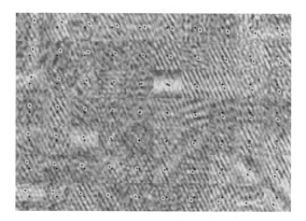

related by a homography from which, using linear methods plus nonlinear optimization, the associated homography matrix can be computed [9].

In Fig. 2.3 a real set of spectral features is shown, where the black crosses represent each patch center and the yellow circles represent the output of PCM. It is important to note that the number, size, and position of spectral features are set beforehand: therefore, neither a search nor a correspondence process needs to be performed.

2.4 Implementation and Results

Summarizing, Algorithm 1 shows the proposed procedure to estimate the position and orientation, Algorithm 2 shows the procedure to determine the displacement between patches, and in Algorithm 3 the homography decomposition process is detailed.

Algorithm 1 Position and orientation estimation: function POSEESTIMATION(i_t, i_{t-1})

Extract patches p_{it} and p_{it-1} from i_t y i_{t-1}
for $\forall\{p_{it}, p_{it-1}\}$ **do**
  $\Delta\mathbf{d}_i \leftarrow$ FINDDISPLACEMENT(p_{it}, p_{it-1})
  $\mathbf{x}_{it} \leftarrow \mathbf{x}_{it-1} + \Delta\mathbf{d}_i$
end for
$H_\lambda \leftarrow$ FINDHOMOGRAPHY($\mathbf{x}_{it}, \mathbf{x}_{it-1}$)
$R, \mathbf{n}, \mathbf{T}/d \leftarrow$ GETRTN(H_λ)
return $R, \mathbf{n}, \mathbf{T}/d$

Algorithm 2 Patches displacement determination: function FINDDISPLACEMENT (p_{it}, p_{it-1})

$P_{it} \leftarrow \text{FASTFOURIERTRANSFORM}(p_{it})$
$P_{it-1} \leftarrow \text{FASTFOURIERTRANSFORM}(p_{it-1})$
$C \leftarrow \text{CROSSPOWERSPECTRUM}(P_{it}, P_{it-1})$
$r \leftarrow \text{INVERSEFASTFOURIERTRANSFORM}(c)$
$\Delta\mathbf{d}_i \leftarrow \text{argmax } r$
return $\Delta\mathbf{d}_i$

Algorithm 3 Homography matrix decomposition: function GETRTN(H_λ)

$U_\lambda, \Sigma_\lambda, V_\lambda^T \leftarrow \text{SVDecomp}(H_\lambda)$
$H \leftarrow H\lambda/\sigma_2$
$U, \Sigma, V^T \leftarrow \text{SVDecomp}(H)$
$\begin{bmatrix} \mathbf{v}_1 & \mathbf{v}_2 & \mathbf{v}_3 \end{bmatrix} \leftarrow V$

$$\mathbf{u}_1 \leftarrow \frac{\mathbf{v}_1\sqrt{1-\sigma_3^2} + \mathbf{v}_3\sqrt{\sigma_1^2-1}}{\sqrt{\sigma_1^2-\sigma_3^2}} \quad ; \qquad \mathbf{u}_2 \leftarrow \frac{\mathbf{v}_1\sqrt{1-\sigma_3^2} - \mathbf{v}_3\sqrt{\sigma_1^2-1}}{\sqrt{\sigma_1^2-\sigma_3^2}}$$

$\mathbf{n}_1 \leftarrow \mathbf{v}_2 \times \mathbf{u}_1 \quad ; \qquad\qquad\qquad \mathbf{n}_2 \leftarrow \mathbf{v}_2 \times \mathbf{u}_2$
Choose only the two physically possible solutions (this ensures that $\mathbf{n}_1$ and $\mathbf{n}_2$ have n_z positive component)
$U_1 \leftarrow \begin{bmatrix} \mathbf{v}_2 & \mathbf{u}_1 & \mathbf{n}_1 \end{bmatrix} \quad ; \qquad\qquad U_2 \leftarrow \begin{bmatrix} \mathbf{v}_2 & \mathbf{u}_2 & \mathbf{n}_2 \end{bmatrix}$
$W_1 \leftarrow \begin{bmatrix} H\mathbf{v}_2 & H\mathbf{u}_1 & H\mathbf{v}_2 \times H\mathbf{u}_1 \end{bmatrix} \quad ; \qquad W_2 \leftarrow \begin{bmatrix} H\mathbf{v}_2 & H\mathbf{u}_2 & H\mathbf{v}_2 \times H\mathbf{u}_2 \end{bmatrix}$
$R_1 \leftarrow W_1 U_1^T \quad ; \qquad\qquad\qquad\qquad R_2 \leftarrow W_2 U_2^T$
$T_1/d \leftarrow (H - R_1)\mathbf{n}_1 \quad ; \qquad\qquad T_2/d \leftarrow (H - R_2)\mathbf{n}_2$
Choose the solution with n_z of each normal plane vector closest to zero
return $R, \mathbf{n}, \mathbf{T}/d$

2.4.1 Spectral Features Evaluation

In order to evaluate the performance of the spectral features in comparison with the intensity features, we use Shi-Tomasi algorithm [15] to detect intensity features in the first frame and Lucas–Kanade algorithm [4] to track these features in the second frame. OpenCV implementations of these algorithms are called `goodFeaturesToTrack()` and `calcOpticalFlowPyrLK()`. The evaluation was done using a camera mounted on a conveyor belt, shown in Fig. 2.4a, simulating a camera movement along Y axis at a constant height. In this way two frames differ only on a pure translation, without changes in scale or angles that affect the test. The displacement of the conveyor belt is measured with a laser telemeter, and the running distance in all the tests is of 0.3 m. The parameters estimated using spectral features are plotted in red, and those estimated using optical flow are plotted in blue. The texture of the floor seen by the camera is shown on Fig. 2.4b.

The performance of the algorithm with both types of features is tested using a zone in the conveyor belt plenty of corner-like features. In this case both approaches perform with low error and high stability. The results are shown in Fig. 2.5: the

(a)   **(b)**

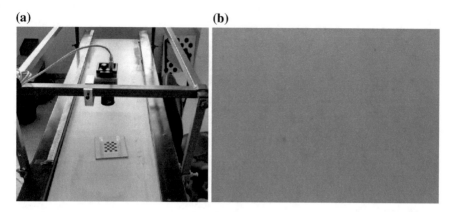

Fig. 2.4 a Camera mounting over a conveyor belt used to compare the performance of spectral feature against Shi-Tomasi algorithm. **b** Floor texture

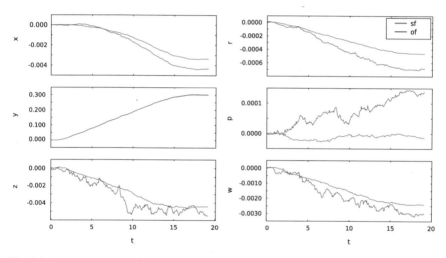

Fig. 2.5 Pose estimation using a textured floor. Plots are x, y, z in m and roll, pitch, yaw in *rad* versus time in s. *Red* spectral features, *Blue* corner-like features

distance measurements along the Y, X and Z axes, and the calculated yaw, pitch and roll angles using both types of features.

In Fig. 2.6 the estimated odometry using the conveyor belt texture with less corner-like features is shown, where the estimation with spectral features are plotted in red and the remaining in blue. As can be seen, the measurements calculated using spectral features are more accurate and stable.

Figure 2.7 shows a situation (pretty common when the floor contains low quality of corner-like features) where the intensity features failed, making the computation of the odometry totally incorrect. This failure is a consequence of a mismatch in the correlation of features, and occurs even more when the image goes out of focus, which

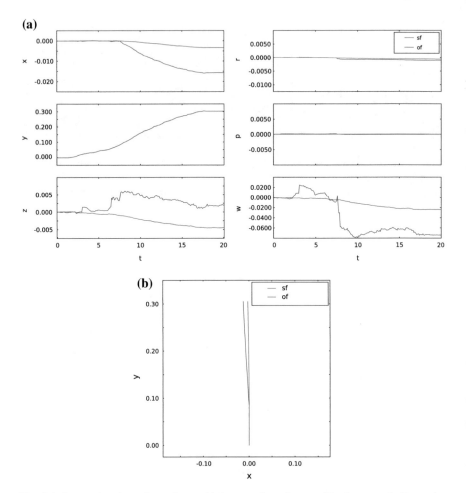

Fig. 2.6 Pose estimation using a floor with low number of corner-like features, similar to that shown in Fig. 2.4a. *Red* spectral features, *Blue* corner-like features

is a very usual situation during a quadrotor flight. In the third image of the sequence shown in Fig. 2.8 it is possible to appreciate this mismatch on the correlation of the features used by the optical flow algorithm, which are drawn in blue. This sequence corresponds to the pose estimation shown in Fig. 2.7d.

2.4.2 Pose Estimation in Simulated Quadcopter

The evaluation of the proposed visual pose estimation approach is performed with synthetic images obtained from a simulated quadrotor. In order to generate a six

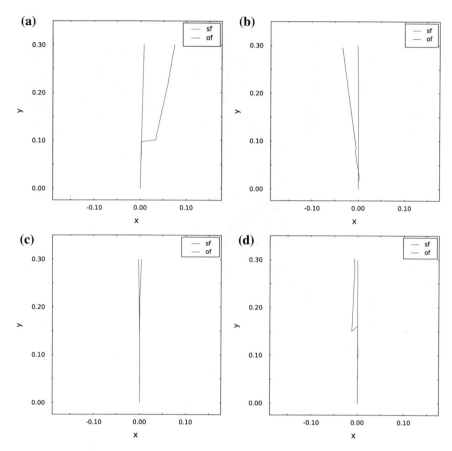

Fig. 2.7 *XY* plot of the pose estimation in a floor with low corner-like features. *Red* spectral features, *Blue* corner-like features

degrees of freedom motion similar to the motion of a real quadrotor, a simulated dynamic model was used. The truth robot position and orientation obtained in this way are then used to crop a sequence of images from a big one representing the observed flat surface. The ground truth pose is also used for evaluation purposes. The simulation of the quadrotor is based on Simulink, and the dynamic model is presented in [6]. Figure 2.9 shows an example of the path followed by the quadrotor used to generate the synthetic dataset.

The path consists on a change of altitude followed by two loops maintaining constant radius. During the loops, the heading angle, also called yaw angle, was set to grow up to 2π radians.

The images were obtained from a *virtual* downward-looking camera following the path described above, cutting portions of 640×480 from a bigger image of uniformly distributed noise in order to simulate a carpet. The virtual camera was

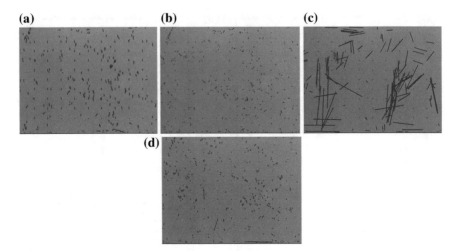

Fig. 2.8 Image sequence corresponding to a wrong pose estimation using intensity features. The floor texture is a low quality corner-like features type, similar to that shown in Fig. 2.4a. *Red* spectral features, *Blue* corner-like features

Fig. 2.9 Simulated position of a quadrotor with a six-degrees-of-freedom motion

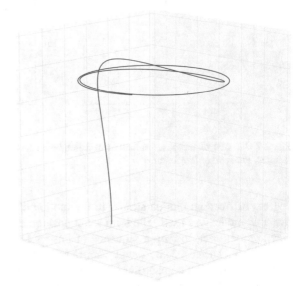

configured with a pixel size of 5.6 μm and a focal length of approximately 1 mm. The algorithm was set with 42 patches of 128 × 128 pixels, equally distributed in the image.

Fig. 2.10 Estimation of the XY-position and yaw angle of the UAV during a 20 s flight

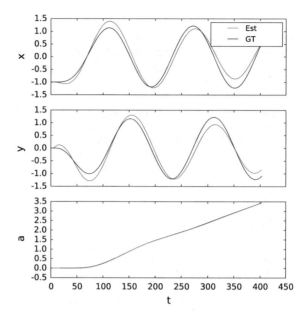

In Fig. 2.10 the estimated parameters together with the ground truth are shown. The graphic at the top shows the X position estimation of the UAV, which performs a total of 2.5 m of change in the complete trajectory. The Y position estimation is plotted in the middle, and it has a similar behavior to the X one. As can be seen, the estimation error remains bounded in both axes all the time. The last graphic shows the yaw angle estimation, which follows the ground truth with a very small error.

2.5 Conclusions

In this work a new approach for visual estimation of the pose change of a quadrotor with a down-looking camera was presented. The proposed algorithm is based on the plane-induced homography that relates two views of the floor, and uses what we call "spectral features" to establish point-correspondences between images.

The main advantage of using spectral features as in this implementation is its robustness in low quality corner-like features floors. Evaluation of this was done using a conveyor belt to simulate a displacement of the camera, and comparing the performance of the spectral features with the Shi-Tomasi intensity features. The spectral features have shown to be more accurate and stable than the intensity features, especially in those scenarios with low quality corner-like features which appears frequently when the camera goes out of focus.

The evaluation of the visual algorithm using a synthetic dataset has shown that the XY-position is estimated without significant absolute error, despite the typical accumulated error of the integration process. It is important to note that the view changes introduced by the orientation change (roll and pitch) over the flight did not induce any considerable error in the XY-position estimation. Likewise, the estimation of the heading (yaw) angle has shown to be accurate enough to be used in an IMU-camera fusion schema.

Acknowledgments This work was partially funded by the Argentinean institutions Universidad Tecnológica Nacional through the project 'Fusión Sensorial para Estimación de Posición y Orientación 3D", UTN-PID-2155, and the National Agency for Science and Technology Promotion through the project "Autonomous Vehicle Guidance Fusing Low-cost GPS and other Sensors", PICT-PRH-2009-0136, both currently under development at CIII, UTN, Córdoba, Argentina.

References

1. Angermann M, Frassl M, Doniec M, Julian B, Robertson P (2012) Characterization of the indoor magnetic field for applications in localization and mapping. In: 2012 International Conference on Indoor Positioning and Indoor Navigation (IPIN), pp. 1–9
2. Araguás G, Sánchez J, Canali L (2010) Monocular visual odometry using features in the fourier domain. In: VI Jornadas Argentinas de Robótica. Instituto Tecnológico de Buenos Aires, Buenos Aires, Argentina
3. Araguás G, Paz C, Gaydou D, Paina GP (2014) Quaternion-based orientation estimation fusing a camera and inertial sensors for a hovering UAV. J Intell RobotSyst 77(1):37–53
4. Baker S, Matthews I (2004) Lucas-kanade 20 years on: a unifying framework: Part 1: the quantity approximated, the warp update rule, and the gradient descent approximation. Int J Comput Vis 56(3):221–255
5. Bonin-Font F, Ortiz A, Oliver G (2008) Visual navigation for mobile robots: a survey. J Intell Robot Syst 53(3):263–296
6. Corke P (2011) Robotics, vision and control, springer tracts in advanced robotics, vol 73. Springer, Berlin
7. Faugeras O, Luong QT (2004) The geometry of multiple images: the laws that govern the formation of multiple images of a scene and some of their applications. MIT press, Cambridge
8. Fischler MA, Bolles RC (1981) Random sample consensus: a paradigm for model fitting with applications to image analysis and automated cartography. Commun ACM 24(6):381–395
9. Hartley R, Zisserman A (2003) Multiple view geometry in computer vision. Cambridge University Press, Cambridge
10. Kaminski JY, Shashua A (2004) Multiple view geometry of general algebraic curves. Int J Comput Vis 56(3):195–219
11. Kuglin CD, Hines DC (1975) The phase correlation image alignment method. Proc Int Conf Cybern Soc 4:163–165
12. Li B, Gallagher T, Dempster A, Rizos C (2012) How feasible is the use of magnetic field alone for indoor positioning? In: 2012 International Conference on Indoor Positioning and Indoor Navigation (IPIN), pp. 1–9
13. Ma Y, Soatto S, Kosecká J, Sastry SS (2010) An invitation to 3-d vision: from images to geometric models. Springer, New York
14. Scaramuzza D, Achtelik M, Doitsidis L, Friedrich F, Kosmatopoulos E, Martinelli A, Achtelik M, Chli M, Chatzichristofis S, Kneip L, Gurdan D, Heng L, Lee GH, Lynen S, Pollefeys M, Renzaglia A, Siegwart R, Stumpf J, Tanskanen P, Troiani C, Weiss S, Meier L (2014) Vision-controlled micro flying robots: from system design to autonomous navigation and mapping in GPS-Denied environments. IEEE Robot Autom Mag 21(3):26–40

15. Shi J, Tomasi C (1994) Good features to track. In: 1994 IEEE computer society conference on computer vision and pattern recognition, 1994. Proceedings CVPR'94, pp. 593–600
16. Weiss S, Achtelik MW, Lynen S, Achtelik MC, Kneip L, Chli M, Siegwart R (2013) Monocular vision for long-term micro aerial vehicle state estimation: a compendium. J Field Robot 30(5):803–831
17. Zitová B, Flusser J (2003) Image registration methods: a survey. Image Vis Comput 21(11):977–1000

Chapter 3
Simultaneous Navigation and Mapping in an Autonomous Vehicle Based on Fuzzy Logic

Álvaro Luiz Sordi Filho, Leonardo Presoto de Oliveira, André Schneider de Oliveira, João Alberto Fabro and Marco Aurélio Wehrmeister

This research presents the navigation control and mapping of an autonomous car by fuzzy logic that enables automatic obstacle avoidance in unknown environments. The strategy is based on a map of the environment, which is created according to navigation, to plan the trajectories avoiding obstacles through the search algorithm A*. The proposed approach is evaluated in a virtual environment, where the autonomous car should move among different obstacles.

3.1 Introduction

Autonomy is the capability of a vehicle (or robot) to move around a known environment, partially known or unknown, based on the perceptions of the environment(sensing), by building the map (mapping), making it possible to plan and replan the routes to the destination point, maneuvering around the obstacles without any interference from an outer source.

Á.L. Sordi Filho (✉) · L.P. de Oliveira · A.S. de Oliveira · J.A. Fabro · M.A. Wehrmeister
Graduate Program on Applied Computing (PPGCA), Federal University
of Technology-Paraná (UTFPR), Curitiba, Brazil
e-mail: alalvaro_7@hotmail.com

L.P. de Oliveira
e-mail: leonardopoliveira@hotmail.com

A.S. de Oliveira
e-mail: andreoliveira@utfpr.edu.br

J.A. Fabro
e-mail: fabro@utfpr.edu.br

M.A. Wehrmeister
e-mail: wehrmeister@utfpr.edu.br

© Springer International Publishing Switzerland 2017
N. Nedjah et al. (eds.), *Designing with Computational Intelligence*,
Studies in Computational Intelligence 664,
DOI 10.1007/978-3-319-44735-3_3

The fuzzy logic is a tool used in the development of control strategies that allows a higher level of flexibility in the rules of the controller, and so, allows the implementation of the autonomous capability. The operation mode of a fuzzy system allows different approaches, like in [1], where the implementation of a fuzzy logic is used in reconfigurable systems, used in the speed in a vehicle's cruise control. Another similar approach is found in [2]. The fuzzy logic can also be used to control the motors in an electric vehicle as showed in [3].

Strategies for ADAS (advanced driving assistant systems) are also a reasonable application for fuzzy systems. In [4], a Fuzzy approach to adjust a PID control is discussed, to guide the vehicle to keep inside the lane. In [5], approach for collision avoidance for autonomous vehicle is taken. In [6] fuzzy systems are used to control the speed of a vehicle's cruise control, with online learning.

The navigation in a dynamic environment, as roads and highways, is a complex task due the amount of obstacles and environment changes that may happen (failures in the road, lack of traffic signs, inter-vehicle interaction, animals, pedestrian crossing, etc). The application of fuzzy systems is advised to situations like these, in [7], for example, a method to navigate in unknown environments based on different type of sensors is introduced. In [8], fuzzy systems are utilized in a autonomous vehicle's distributed control system.

This work focus on the capability of a vehicle to be autonomous, which means nothing more than an agent that is able to extract the information of an environment and use it to move around such environment in an efficient manner. The possible applications for these robots is huge, they can be used in the industry (AGV - autonomous guided vehicles), in the military (UAV - unguided autonomous vehicle), exploration, among others [9].

This work proposes a fuzzy system to control the navigation of an autonomous vehicle inserted in and static, partially observable (the sensing system cannot access all the environment information at all times), deterministic, discreet, and single agent. The validation experiments are developed in a simulation environment, and for these, a realistic vehicle model with sensors and actuators available in the market were used.

In the following sections, the concepts and the techniques used in the simulation are going to be discussed (Sects. 3.2, 3.3 and 3.4), the methodology adopted in the development (Sect. 3.5), the results and the final conclusions (Sects. 3.6 and 3.7).

3.2 Virtual Robot Experimentation Platform

The VREP (virtual robot experimentation platform) has resources to create, compose, and simulate robots. It has verification systems and can monitor remotely the actions performed by the robot under review [VREP 2015]. The VREP calculation module can determine the optimal parameters of a mobile joint to achieve the correct positioning, quickly calculates the possibility of a collision, allows planning a way to run in finite space, operates and interacts with programmed mechanisms and scene objects.

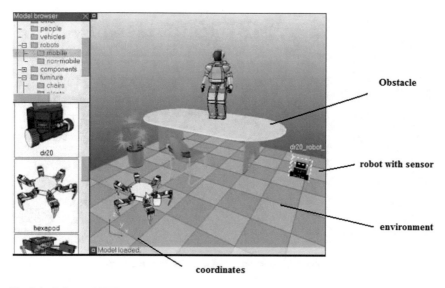

Fig. 3.1 Software VREP

The scene's objects can have cameras and lights, proximity sensors, force sensors, image interpretation cameras, and paths defined on 2D and 3D graphics. The simulation can be started, paused, and stopped at any instant of time and can be performed in real time or by approximate running time. Figure 3.1 illustrates the simulation environment VREP and scene's objects.

The self-guided car was imported and prepared from the robot library available in VREP. Environmental obstacles were inserted with properties "collidable," "detectable," and "measurable". The choice of the path to be traveled by the vehicle considered a starting point, the properties of the chosen path, coordinate or objective to achieve (meaning the goal location). Data was also inserted in the environment as the minimum and maximum detection distance and the minimum diameter of carriage movement.

The VREP has the application programming interface (API) that lets you access your library and services available through another programming language or application. It is possible to use C/C++, Java, Python, or mathematical software Matlab.

3.3 Autonomous Vehicle

Autonomous vehicles can have different topologies, which are dependent mainly of the mobility and maneuvering necessary for the application. An autonomous car can be considered as a mobile robot with different sensing systems, used in the perception of the environment. This robot is also able to move without the interference of an

Fig. 3.2 Renault Fluence autonomous vehicle, and range of the sensing systems

extern operator. This vehicle uses a drive system called Ackerman geometry, which hold four wheels, of which two are fixed and the other two are directional. A further discussion about this geometry is available in [10].

The environment perception is acquired by sensors that measure the environmental quantities. Autonomous vehicle, typically, use sensors to identify objects (cameras, radars, LIDAR-light detection, and ranging, among others). In this context, the current work utilizes a regular sedan vehicle, more specifically a Renault Fluence, which was modeled in a virtual experimentation system V-REP (Virtual Robot Experimentation Platform), of Coppelia Robotics [11]. In the vehicle, were placed four radar systems (ultrasound) of medium range to identify possible objects on the sides (typically used to lane change assistance, and automated parking). At the front, a long range radar (ultrasound) was placed to detect object even in high speed. At the rear, parking sensors (ultrasound) was placed and also a LIDAR sensor (Hukuyo, a laser sensor) to help in the environment perception, as can be seen in Fig. 3.2.

3.4 Navigation and Mapping

The Renault Fluence's autonomy is achieved with a group of techniques called SLAM (simultaneous localization and mapping), which allows that the building of the map itself, is done during the robot's navigation (a further discussion about this approach is presented in [12]).

SLAM is one of the most addressed methods in the robotic area. This technique refers to how the robot is going to behave in a unknown environment and how it

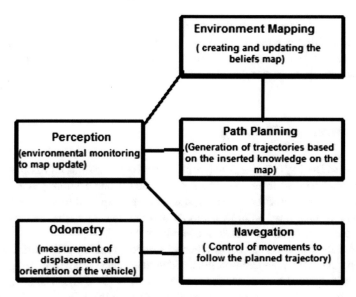

Fig. 3.3 Simultaneous localization and mapping structure (SLAM)

is going to develop the knowledge, while it moves around in the environment. The environment perception is performed by sensors that can be classified accordingly to its measurement as: proprioceptives (measure the greatness of the vehicle's inside) or exteroceptives (measure environmental greatness). They can also be classified as: passive (measure the environment energy) or active (they emit their own energy) [12]. There sensors detect the information that the robot needs to create its own navigation map. Figure 3.3 represents the inner structure of a SLAM system.

Sensors capture the perceptions of the environment and send them to the Mapping mechanism, which is responsible to create and update the knowledge map. The environment information allows the robot to locate itself (inside the environment), and then deliberate on how to act (navigate). This structure can also predicts the presence of an object that is not currently present in the environment.

The constant updates on the map is a basic need of the SLAM technique, because as the robot is moving around the map and detecting new obstacles on the map, in a second pass through the same position the object might not be there anymore. Is this case, if a static map was used, the robot will act as if the object was still there, resulting in errors.

The creation of the knowledge map uses a data structure called costmap, in which the robot define the cost of each point in the map. The costmap is used to calculate the route to the destination (trajectory planning), which is going to be used as reference to the navigation controller. The trajectory is calculated using the costmap, looking for the path with the smallest cost [13].

The information added to the map have error in both position and size, due to the face of measurement errors, which directly interfere in the navigation and might lead to collisions. Therefore, every object present in the map has a shadowed boundary, with costs smaller than the obstacle itself, which lower the probability of the robot choosing that region on its route. This a fundamental characteristic in situations where the vehicle must calculate if it can move through a narrow pass.

3.4.1   Trajectory Planning

According to Delling [14], the A* algorithm search for and finds if possible, the path with the smallest cost from a start to a goal node (of one or more possible goals). To achieve this A* algorithm move through a graph (map) and follows the way with the lower expected cost, holding a priority ordered queue, in which each elements represents a piece of an alternate route along the way.

The A* algorithm uses a cost function of a node n (usually called f(n)), which is defined by the cost already known, g(n), added to the cost of the estimated path, h(n) (h for heuristic), to choose in which order each node is going to be visited in the tree. This cost function is the sum of two functions: $f(n) = g(n) + h(n)$.

Delling [14] also says that h(n) must be na admissible heuristic, it should not overestimate the cost for the destination. Therefore, for the application that must calculate the distance between two points, h(n) can represent the straight line between start and objective, because this is always the smallest distance between two points or nodes. This is also called the Euclidean distance.

If the heuristic h satisfies the additional condition $h(x) = d(x, y) + h(y)$ for each edge (x, y) of the graph (where d indicates the length of edge), h is the called consistent. In this case, the A* algorithm can be implemented more efficiently, roughly, no node needs to be processed more than once and A* is equivalent to running Dijkstra's algorithm with reduced cost $d'(x, y) = d(x, y) - h(x) + h(y)$ [10].

The A* algorithm is responsible for allowing the autonomous vehicle plan the optimal path between the point that it is up to the chosen goal. An adjustment was made in the A* algorithm that instead of returning all points of the optimal path, the A* algorithm returns only the conversion points present in the optimal path. It was determined that conversion point is a point at which the car should make a turn, the distance between two turning point should be done in a straight line. For example, if the car is in the space (0, 0) (x, y); and must get to the point (10, 10). The A* algorithm can return the following way: (0.1) (0.2) (0.3) (0.4) (0.5) (0.6) (0.7) (0.8) (0.9) (0.10) (1.10) (2.10) (2.10) (4.10) (5.10) (6.10) (7.10) (8.10) (9.10) (10.10); in all are 20 points. With the adjustment made the algorithm returns only (0, 0), (0, 10), (10, 10). In this way, you can determine a straight line between the points (0,0) and (0.10), and another straight line between the points (0.10), (10:10), and this will be the path that the car will follow.

3.4.2 Navigation

Fuzzy systems are based on fuzzy logic, which has different characteristics from traditional logic. A traditional logical proposition has two specific situations: it is completely true or entirely false. But in fuzzy logic, a premise can vary between true or false, that is, instead of being only values 0 or 1, can occur in a premise assume a value between 0 and 1. As can happen to a premise be partly true or false [15].

The process of a fuzzy system is based on fuzzy sets, the membership functions, and the fuzzy rules. The sets are partitions of all the possible values for the input variables. The membership functions define the fuzzy sets. The rules are used in the system inference engine. They use complementary operators, union, and intersection to establish the relationship between the input variables and system output [Coppin 2013].

The partition of a fuzzy set assigns degrees of relevance to the elements of this set. The data used to create the rules were generated based on tests done on the platform. Aspects were tested as braking and acceleration time of the simulated vehicle. From these tests it was created a data table, which served as the entrance to the creation of inference rules. To create the fuzzy inference rules were used the concepts of Wang Mendel algorithm [16].

This Wang Mendel algorithm is used for the automatic extraction of the relevant rules to a fuzzy system and can be summarized in the following steps: Define the number of linguistic terms and partition the universe of all the input variables; Building a fuzzy rule for each member of the set of training points—For each input variable, select the higher level membership function; Calculate the degree of activation of all rules using an appropriate operator.

In the autonomous vehicle, the fuzzy system is designed to control not only the car's speed, but also control the rotation of the wheels when it is needed. In all three are applied Fuzzy controllers.

The first fuzzy system consists of two inputs (distance and speed of the front wheels) and an output (speed). The goal of this fuzzy system is to ensure that the car do not collide with the obstacle. Figure 3.4 illustrates this system. The variable "distance" ranges between 5 and 20 m, the "rotation" variable ranges between −40° (left oriented) and 40° (right oriented) and the output ranges between 15 and 40 km/h. Another simple control system complements the operation and ensures that if the car is 0.5 m from the obstacle, a reverse maneuver is performed, since from that position this is the only alternative.

When the car is at a shorter distance than 5 m, the control is based on the front sensors and causes the car to turn to the side closest to the desired point. This control determines the difference between the direction the car is and the direction it should be to follow the optimal path. The output is the angle that the car should turn to follow the desired path and varies from 180 (left) to 180 (right), as illustrated in Fig. 3.5.

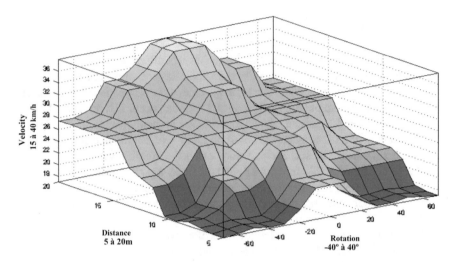

Fig. 3.4 Auxiliary control system that helps to avoid car collision with environmental obstacles

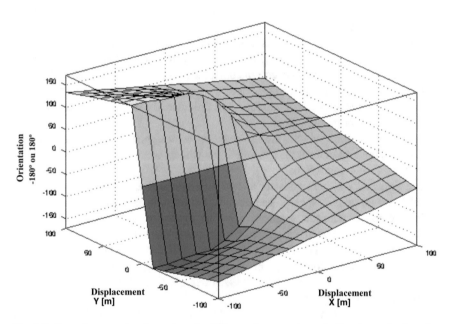

Fig. 3.5 Control system that determines the difference between the car's direction and orientation of the goal point

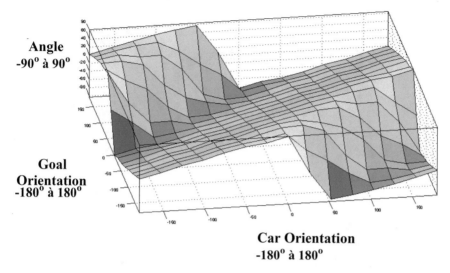

Fig. 3.6 Fuzzy system to supervise the car turning, so that it reaches the correct orientation to reach the goal

The third fuzzy control system (Fig. 3.6) oversees the car turning and determines both rotation of the wheels, and the speed of movement. This control is important because if the car is in a narrow place, the maneuver cannot be performed in a single movement.

3.4.3 Decision Tree

Decision tree is the simplest form among the most used decision models, but is quite effective. Among its main strengths, its transparency and the ease of developing are highlighted.

The decision tree structure is very similar to the if-then structure, widely used in expert systems and rating systems [17].

At the entrance of a decision tree are received attributes that can be continuous or discrete, then the tree will reach a final decision based on their tests. In the tree structure, each node represents a different test and each leaf node of this tree represents a value that is returned if this leaf node is reached.

In the decision tree, each node is the knowledge of the expert leading the search for one of its child nodes. So as tou move down the tree, the desired system configuration will be selected and thus choosing the desired behavior [18].

For the autonomous vehicle, the decision tree is used to select which group of actions the agent must take, i.e., to decide if it should just follow the path determined by the trajectory planner, dodge an obstacle, or to stop and make a reverse in case of the available space is not sufficient to maneuver.

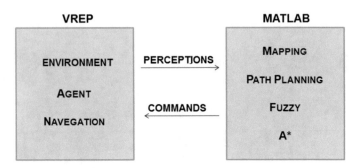

Fig. 3.7 Architecture simulation

3.5 Simulation

Figure 3.7 presents the architecture of the simulation to be developed.

The VREP is responsible for simulating the environment, the agent (vehicle), and the navigation positioning information (which is calculated by MATLAB). Agent perceptions (signals picked up by the sensors) are sent from VREP to MATLAB and this in turn is responsible for performing the trajectory calculations (A*), determine the correct speed that the car should be in accordance with the current situation (Fuzzy), in addition to building the mapping and planning trajectories. MATLAB then returns the commands, which are these trajectory and velocity information that the vehicle must follow.

For the simulation to work correctly, there must be integration of code developed in MATLAB, with the components created in VREP. Communication between the two software is executed via socket, and to facilitate this, in this paper we chose to use the API developed by Coppelia (company responsible for VREP).

The modeling agents is described in more detail in the online tutorial VREP; however, it is required that each vehicle component is declared as a clear object name, since to access these components via MATLAB the names that were defined in VREP are used as unique identifiers.

For example, when you want to turn the front wheel 30° right, the following command is sent.

- vrep.simxSetJointTargetPosition(clientID, RhtWheelHandle, −degtorad(30), vrep.simx_opmode_oneshot);

The vrep.simxSetJointTargetPosition command is used to rotate a particular component present in VREP. Arguments are, respectively, clientID − name of the simulation; RhtWheelHandle − the name given to the right front wheel; −degtorad(30), which converts 30° to its equivalent value in radians, and vrep.simx_opmode_oneshot − representing that this command must be run only once. In this example, it is clear that the choice of name facilitated in performing the function, as RhtWheel can be easily associated with the front right wheel.

3.6 Case Study

The case study is a Renault Fluence (Fig. 3.8) vehicle, which is placed in a virtual environment with many obstacles. As defined by the Ackerman geometry, only the front wheels determine the direction, i.e., the rear wheels are free. When the car needs to make a turn, only the front wheels must be acted on. The car's size is about 4 m long and 2 m wide.

The environment is a closed area of 100 m × 100 m, as illustrated in Fig. 3.9. In this space were defined spaces by which the vehicle can follow and also obstacles that it should divert. Altogether there are four obstacles in the shaped of cubes of varying sizes. Besides the obstacles, there are walls which limit the passage of the vehicle.

The side tracks (bounded by walls) are narrow to hinder a possible curve the car must do. This difficulty requires the control to be more efficient because, depending on the speed it is not possible that the car maneuver without having to move in the opposite direction (reverse) or maneuver itself to fir the curve.

The experiment starts with the car stopped at a certain point and the map of empty knowledge. At that moment, the objective point is designated for the autonomous vehicle. The first task is to carry out the planning of the trajectory based on the knowledge map, then the path is calculated by A* and then when the algorithm returns the "sub-goals" (turning points) the car starts moving at the resulted path. The fuzzy controllers are responsible for making the car follow the planned trajectory.

Fig. 3.8 Autonomous vehicle—virutal Renault Fluence

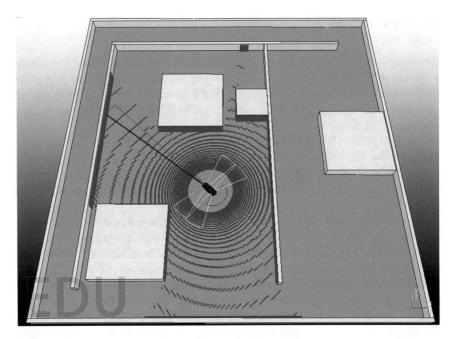

Fig. 3.9 A car that interact in this environment, which is unknown at first (but is being mapped), whose object reaches a certain point, traversing the optimum path (derived from the A* algorithm)

The car follows the optimum path at the same time it updates the environment map. When one eventually is very near or in the planned trajectory, the fuzzy control action operates on the front sensor for the vehicle to deviate from obstacles.

Figure 3.10 depicts the map of the environment being built as the car moves. The brighter areas represents he position where the obstacle was encountered, while the area in gray is the inflation that the car has to calculate around obstacles.

When the obstacle is overcome, the control indicates that the car should go back to follow the path determined by the trajectory planner, but the car may not be oriented in a way that is must only move forward. At this point, another fuzzy controller is activated to correct the orientation of the vehicle according to the desired trajectory (determined by the trajectory planner).

After adjusting the orientation of the vehicle to the optimum path, the car return to the state of following the path determined by the trajectory planner to reach its goal. Due to the car's size, it was determined that the goal is any point in a distance of 1 m from the goal point (tolerance range). This measure was taken to prevent the vehicle to keep making small maneuvers to be exactly at the desired point, since these maneuvers took a long time and are negligible, since the car was already above the point and was just trying to align the car's center with the goal point.

Fig. 3.10 Environment partially mapped by the vehicles with help of the perception system

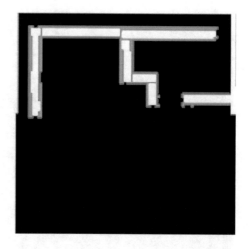

The fuzzy controller for steering correction, used interactively, enables the direction (orientation) fine tuning. The fuzzy controller responsible for the vehicle's speed depending on the distance of the obstacles was important to ensure that the car does not collide with objects.

The correction in the direction of the vehicle is an intrinsic difficulty to the problem, because as it is a regular vehicle, any direction adjustment requires continuous forward and backward movements, until it reaches the desired angle.

The vehicle was also tested in a different scenario. This time the scenario is a bit closer to a real-life situation. The car was placed inside a parking lot of a shopping mall. Figure 3.11 represents this scenario with an upper view. The whole scenario is composed by the shopping mall at the middle, the vehicle at the left upper corner and eight light posts, being 4 at the north and 4 at the south wall The textures of the shopping mall exterior and the parking lot floor were removed to improve performance of the simulation.

As the previous experiment, different positions were used as goal for the vehicle. When the car reached its destination, another point was set as goal. By switching destinations, it was nearly possible to map the entire parking lot with the shopping mall at the middle. Figure 3.12 contains the resulting map of the scene.

The places that were not mapped in Fig. 3.12 are spots where the vehicle did not get close enough so its sensors were able to identify the walls. This happened because during the path planning stage, the search algorithm was able to find a better path that did not go through the unmapped spots.

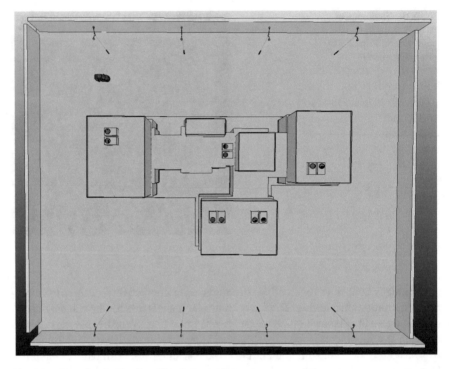

Fig. 3.11 Shopping mall and parking lot scenario

Fig. 3.12 Partial map of the shopping mall and parking lot scenario

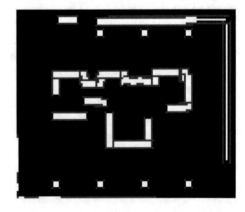

3.7 Conclusions

The simulations show that the system responds effectively. Initially in a totally unfamiliar environment, with a goal position, the robot calculates the optimal route, this being a straight line. As the robot detects objects and updates its knowledge map, as soon as it finds an obstacle in its path the route is recalculated.

The method A* was used, despite producing the desired result, it requires relatively much computational time, causing the car, in the simulation, stand still while the trajectory planner find the optimal path. The fuzzy control systems (Fuzzy) have proven their efficiency in relation to what was expected. The guidance control to the next sub-goal returned valid responses to any situation where the car and the goal met.

The SLAM used in the work also proved to be satisfactory to map the environment. The detail to be highlighted at this point is that inflation in the mapping area can interfere decisively in the development of history. In the case of this work was necessary to increase the inflation area to ensure that the car does not collide with the corners in the environment. The autonomous vehicle is able to navigate in different environments, with smoother curves, and less narrow lanes.

In addition to different environments, it is suggested to implement a control to make the transition from the direction of the vehicle between softer sub-goals.

Acknowledgments To the Araucaria Foundation and the Renault of Brazil for partial funding from the research project.

References

1. Nedjah N, Sandres PRSS, de Macedo Mourelle L (2014) Customizable hardware design of fuzzy controllers applied to autonomous car driving. Expert Syst Appl 41(16):7046–7060
2. Petkovic D, Issa M, Pavlovic ND, Zentner L (2013) Intelligent rotational direction control of passive robotic joint with embedded sensors. Expert Syst Appl 40(4):1265–1273
3. Peng X, Zhe H, Guifang G, Gang X, Binggang C, Zengliang L (2011) Driving and control of torque for direct-wheel-driven electric vehicle with motors in serial. Expert Syst Appl 38(1):80–86
4. Wang Q, Xu S-Z, Xu H-L (2014) A fuzzy control based self-optimizing PID model for autonomous car following on highway. In: International conference on wireless communication and sensor network (WCSN), pp 395–399, 13–14 December 2014
5. Llorca DF, Milanes V, Alonso IP, Gavilan M, Daza IG, Perez J, Sotelo MA (2011) Autonomous pedestrian collision avoidance using a fuzzy steering controller. IEEE Trans Intell Transp Syst 12(2):390–401
6. Onieva E, Godoy J, Villagra J, Milanes V, Perez J (2013) On-line learning of a fuzzy controller for a precise vehicle cruise control system. Expert Syst Appl 40(4):1046–1053
7. Farooq U, Hasan KM, Amar M, Asad MU (2013) Design and implementation of fuzzy logic based autonomous car for navigation in unknown environments. In: International conference on informatics, electronics and vision (ICIEV), pp 1–7, 17–18 May 2013
8. Martinez-Barbera H, Herrero-Perez D (2014) Multilayer distributed intelligent control of an autonomous car. Transp Res Part C: Emerg Technol 39:94–112

9. Terano T, Asai K, Sugeno M (eds) (2014) Applied fuzzy systems. Academic Press, New York
10. Ackermann J, Bunte T, Odenthal D (1999) Advantages of active steering for vehicle dynamics control
11. Rohmer E, Singh SPN, Freese M (2013) V-REP: a versatile and scalable robot simulation framework. In: IEEE/RSJ international conference on intelligent robots and systems (IROS), pp 1321–1326, 3–7 November 2013
12. Siegwart R, Nourbakhsh IR, Scaramuzza D (2011) Introduction to autonomous mobile robots. MIT Press, Cambridge
13. King J, Likhachev M (2009) Efficient cost computation in cost map planning for non-circular robots. In: IEEE/RSJ international conference on intelligent robots and systems. IROS. IEEE
14. Delling D, Sanders P, Schultes D, Wagner D (2009) Engineering route planning algorithms. Algorithmics of large and complex networks. Springer, New York, pp 117–139
15. Melo LG (2011) Sistema Fuzzy Probabilístico Geração Automática de regras e Defuzzificação Bayesiana. Dissertação de Mestrado em Informática Industrial. Universidade Tecnológica Federal do Paraná, Curitiba
16. Wang L-X, Mendel JM (1992) Fuzzy basis functions, universal approximation, and orthogonal least-squares learning. IEEE Trans Neural Netw 3(5):807–814
17. de Soárez PC, Soares MO, Novaes HMD (2014) Modelos de decisão para avaliações econômicas de tecnologias em saúde. Revista Ciência & Saúde Coletiva 19.10
18. Pozzer, CT (2006) Aprendizado por árvores de decisão. 2010, Universidade Federal de Santa Maria, Rio Grande do Sul

Chapter 4
Fully Scalable Parallel Hardware for Wheeled Robot Navigation Using Fuzzy Control

Nadia Nedjah, Paulo Renato S.S. Sandres and Luiza de Macedo Mourelle

Process control is one of the many applications that took advantage of the fuzzy logic. Controllers are usually embedded into the controller device. This chapter aims at presenting the development of a reconfigurable efficient architecture for fuzzy controllers, suitable for embedding. The architecture is parameterizable so it allows the setup and configuration of the controller, so it can be used for various problem applications. An application of fuzzy controllers was implemented and its cost and performance have been evaluated.

4.1 Introduction

Computational system modeling is full of ambiguous situations, wherein the designer cannot decide, with precision, what should be the outcome of the system. In [7], L. Zadeh introduced for the first time the concept of *fuzziness* as opposed to *crispiness* in data sets.

Fuzzy logic and approximate reasoning [6] can be used in system modeling and control as well as data clustering and prediction, to name only few appropriate applications. Furthermore, they can be applied to any discipline such as finance, image

N. Nedjah (✉) · L. de Macedo Mourelle
Department of Electronics Engineering and Telecommunications, State University
of Rio de Janeiro, Rio de Janeiro, Brazil
e-mail: nadia@eng.uerj.br

L. de Macedo Mourelle
e-mail: ldmm@eng.uerj.br

P.R.S.S. Sandres
Department of Systems Engineering and Computation, State University of Rio
de Janeiro, Rio de Janeiro, Brazil
e-mail: paulo.sandres@eng.uerj.br

© Springer International Publishing Switzerland 2017
N. Nedjah et al. (eds.), *Designing with Computational Intelligence*,
Studies in Computational Intelligence 664,
DOI 10.1007/978-3-319-44735-3_4

processing, temperature and pressure control, robot control, among many others. The fuzzy logic is a subject of great interest in scientific circles, but it is still not commonly used in industry, as it should be. Eventually, we found some literature containing practical applications that is being currently used in industry [3, 4].

There are many related works that implemented a fuzzy controller on a FPGA, but most of them present controller designs that are only suitable for a specific application [2, 4]. Mainly, the designs do not use 32-bit floating-point data. The floating-point data representation is crucial for the sensibility of the controller design. In contrast, all the required computation in the proposed controller are performed by a simple precision floating-point coprocessor.

The purpose of the development of a reconfigurable hardware of a *shell* fuzzy controller, which can include any number of inputs and outputs as well as any number of rules, is the possibility of creating a device that can be used more widely and perhaps spread the concept of fuzzy logic in the industrial final products.

This paper is divided into three sections. First, in Sect. 4.2, we introduce briefly some concepts of fuzzy controller, which will be useful to follow the description of the proposed architecture. Then, in Sect. 4.4, we describe thoroughly, the macro-architecture of the fuzzy controller developed. After that, in Sect. 4.5, we give details about the main components included in the macro-architecture. Subsequently, in Sect. 4.3, we present the fuzzy model used to control the navigational process of a wheeled robot. Then, in Sect. 4.6, we show, via the project of a the fuzzy controller presented, that the proposed architecture is functionally operational and promising in terms of cost and performance. Finally, in Sect. 4.7, we draw some conclusions and point out some new direction for the work in progress.

4.2   Fuzzy Controlllers

Fuzzy control, which directly uses fuzzy rules, is the most important and common application of the fuzzy theory [5]. Using a procedure originated by E. Mamdani [3], three steps are followed to design a fuzzy controlled machine:

1. fuzzification or encoding: This step in the fuzzy controller is responsible of encoding the crisp measured values of the system parameter into a fuzzy term using the respective membership functions;
2. inference: This step consists of identifying the subset of fuzzy rules that can be fired, i.e., those with antecedent propositions with truth degree not zero, and draw the adequate fuzzy conclusions;
3. defuzzification or decoding: This is the reverse process of fuzzification. It is responsible of decoding a fuzzy variable and compute its crisp value.

The generic architecture of a fuzzy controller is given in Fig. 4.1. The main components of a fuzzy controller consist of a *knowledge repository*, the *encoder* or *fuzzification* unit, the *decoder* or *defuzzification* unit and the *inference engine*. The

Fig. 4.1 Generic
architecture of fuzzy
controllers

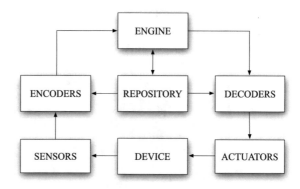

knowledge base stores two kind of data: the fuzzy rules that are required by the inference engine to reach the expected results and knowledge about the fuzzy terms together with their respective membership functions as well as information about the universe of discourse of each fuzzy variable manipulated within the controller. The encoder implements the transformation from crisp to fuzzy and the decoder the transformation from fuzzy to crisp. Of course, the inference engine is the main component of the controller architecture. It implements the approximate reasoning process.

4.3   Fuzzy Models for Wheeled Robot Navigation

The control of a wheeled robot navigation uses a series of control loops to operate on a surface following a predefined trajectory. Figure 4.2 shows the schematics of the used robot.

This application consists of three subcontrollers: *(i)* the steering control, which uses two controllers requiring two inputs and one output, each; the linear and angular speed controls, which use the same control process requiring two inputs and one output. Although this application has four controllers, this paper will only show one of them, because the drivers are identical in pairs, i.e., the membership functions and rules of the controllers are the same for the the linear and angular velocity.

4.4   The Proposed Macro-architecture

The macro-architecture of the proposed fuzzy controller consists of three main units: *(i)* the fuzzification unit (FU), which is responsible for translating the input values of the system into fuzzy terms using the respective membership functions. This unit has as many Fuzzy blocks as required in fuzzy system model that is being implemented,

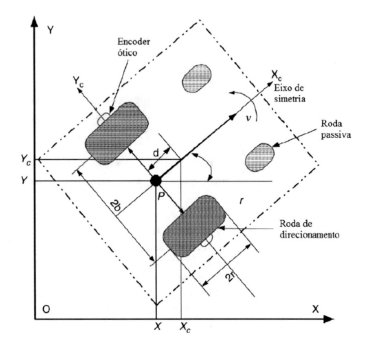

Fig. 4.2 Model of the wheeled robot used in this application [1]

i.e., one for each input variable; *(ii)* the inference unit `Inference`, which checks all the included fuzzy rules, verifying which membership function applies, and if any is so, generating its value and thus identifying the membership functions to be used in the sequel; *(iii)* the defuzzification unit (DU), which is responsible for translating the fuzzy terms back so as to compute the crisp value of the fuzzy controller output. The defuzzification unit includes as many `Defuzzy` blocks as required by the fuzzy system model that is being implemented, i.e., one for each output variable. The block diagram of the proposed macro-architecture is shown in the Fig. 4.3, wherein N and M represent the number of input and output variables, respectively.

Note that, besides the main units, the macro-architecture also includes a component that allows to compute the membership function characteristics, which are used by both the fuzzification and defuzzification units. This component will be called membership function unit (MFU). It includes as many `MF` blocks as required input variable of the fuzzy model. Note that all the membership function-related data are stored in the membership function memory, called MF MEM. This memory is formed by as many memory segments as required input variables, i.e., one for each membership function used. The rules used by the inference unit are stored in a read-only memory block, called `Rules`. Component `Controller`, which in the sequel may be called *main controller*, imposes the necessary sequencing and/or the simultaneity of the required steps of the fuzzy controller via a concurrent finite state machine. More details on this are given subsequently.

The proposed fuzzy controller is designed to be generic and parametric, so it allows configuring the number of input and output variables, the number of linguistic terms used to model the membership functions, and the number of inference rules, so as the fuzzy system model that is being implemented can fit in. Allowing the configuration of these parameters makes it possible, as well as easy, to adjust the controller design to any desired problem.

As it can be seen in Fig. 4.3, at configuration time, all the membership functions used by the controller are computed and stored in the respective MF MEM segment of the membership function memory. All the computed data will be readily available to be used by the pertinent Fuzzy and/or Defuzzy block in the fuzzification and defuzzification unit, respectively. Note that this configuration step is done only once. During the operation step, the fuzzy controller iterates the required steps, triggering the Fuzzy blocks then Inference unit then Defuzzy blocks in sequence. After that, it waits for a new set of input data to be read by the system sensors and thus arrive at the Fuzzy blocks input ports. The finite state machines that control the Fuzzy blocks all run in parallel, so do those that control the Defuzzy blocks.

In the following sections of this chapter, more light will be shed on the internal micro-architecture of the proposed design as well as the control used therein.

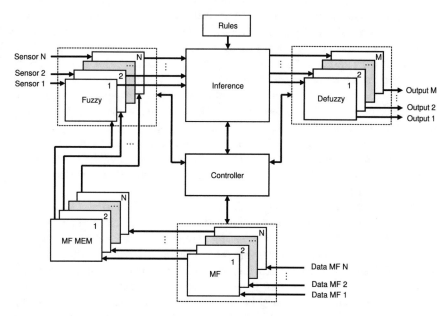

Fig. 4.3 Macro-architecture of the designed fuzzy controller

4.5 Micro-architecture of the Functional Units

In this section, we describe the micro-architecture of the main components, included in the macro-architecture of Fig. 4.3. These are the functional unit responsible for the computation of the member function (MF), including the memory-based component (MF MEM), the basic component responsible for the fuzzification process (Fuzzy), the component that implements the inference process (Inference) using the available rule base (Rules), and the basic component that handles the defuzzification process (Defuzzy). In general, all blocks that perform floating-point computations include an FPU unit, which performs the main mathematical operations with simple precision (32 bits). The operations needed are addition, subtraction, multiplication and division.

4.5.1 Membership Function Unit

A membership function is viewed as a set of linguistic terms, each of which is defined by two straight lines. In the proposed design, the triangular shape is used to represent linguistic terms. Nevertheless, it is possible to adjust the design as to accept other used shapes such as trapezes and sigmoid. Figure 4.4 shows a generic example of membership function with Q linguistic terms, wherein the horizontal axis x represents the controller's input, probably read from a sensor, and the vertical axis y represents the truth degree associated with the linguistic terms. This is a real value, between 0 and 1, handled as a simple precision floating-point number of 32 bits. Linguistic terms of triangular membership function are completely defined by *Max Point* or *mp* and *Range* or *r*, as illustrated Fig. 4.4.

The MF block is designed to compute the values of any variable x, according to $y = ax + b$ of the two straight lines, that represent the linguistic term of the membership function. The required basic data that completely define these shapes need to be identified.

The input data of the MF block are *MaxPoint – Mp*, *Left Interval – Li* and *Right Interval – Ri* for each straight line used to define the linguistic terms of the membership function. The block utilizes them and precompute coefficients a and b accordingly and stored them in the membership function memory segments. Three cases are possible: the leftmost linguistic term (see linguistic term 0 in Fig. 4.4); An in-between linguistic term (see linguistic term 1 and 2 in Fig. 4.4); and finally, the rightmost linguistic term (see linguistic term Q in Fig. 4.4). The computation of a and b of the straight lines of the leftmost, middle, and rightmost linguistic terms are defined as in (4.1)–(4.3), respectively.

$$\mu_L(x) = \begin{cases} 1, & \text{if } x \leq Mp \\ -\frac{1}{Ri} \times x + \frac{Mp}{Ri} + 1, & \text{if } Mp > x \geq Mp + Ri \\ 0, & \text{otherwise} \end{cases} \quad (4.1)$$

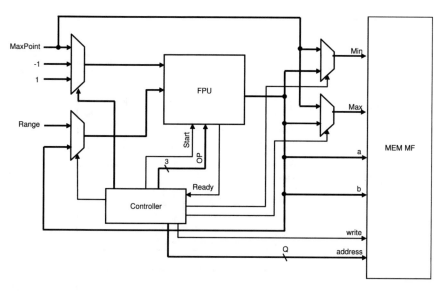

Fig. 4.4 Membership function of Q linguistic terms

$$
\mu_M(x) = \begin{cases} \frac{1}{Li} \times x - \frac{Mp-Li}{Li}, & \text{if } Mp - Li < x \leq Mp \\ -\frac{1}{Ri} \times x + \frac{Mp}{Ri} + 1, & \text{if } Mp > x \geq Mp + Ri \\ 0, & \text{otherwise} \end{cases} \tag{4.2}
$$

$$
\mu_R(x) = \begin{cases} \frac{1}{Li} \times x - \frac{Mp-Li}{Li}, & \text{if } Mp - Li < x \leq Mp \\ 1, & \text{if } x > Mp \\ 0, & \text{otherwise} \end{cases} \tag{4.3}
$$

The micro-architecture of the membership function blocks MF is shown in Fig. 4.5. It uses a floating-point unit to perform the required mathematical operations. The obtained results are then stored in the MF MEM segments.

An MF block includes a controller that is implemented as a finite state machine. It allows to synchronize the setting up of all the linguistic terms, necessary to the complete definition of the membership function for each input variable. The control sequence of this controller is given in Algorithm 1.

4.5.2 Membership Function Memory

As explained earlier, this memory block responds to write commands received from the MF block and read commands issued by the FU. Each word of this memory holds four data that allows the complete computation of the truth degree of a given

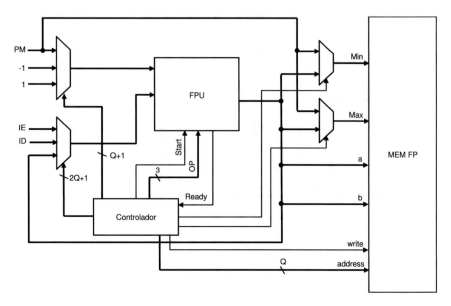

Fig. 4.5 The micro-architecture of the membership function block

linguistic term. The four-fold memory word contains min: minimum limit of the straight line; max: maximum limit of the straight line; a: angular coefficient of the straight line; b: linear coefficient of the straight line.

So, every time the MF block requests a memory write, this memory block register these values at an address, that represents the order number of the line within all the line that need to be processed, starting from zero. This block also allows the configuration of the number of lines that can be registered in the memory, which will depend on parameter Q, which determines the number of linguistic terms per membership function.

4.5.3 Fuzzification Unit

The Fuzzy block performs the necessary computation to obtain the fuzzy version the input value. The computation consists of a comparison that may, in most cases, be followed by a multiplication then an addition, depending on the comparison result. This is repeated Q times for all the linguistic terms included in the membership function of the input variable under consideration. The Fuzzy block micro-architecture is shown in Fig. 4.6. It includes a Comparator that determines in which linguistic term range the input value falls, 2 sets of Q flip-flops to hold the result of the comparison. Their contents identify which linguistic terms are actually active.

Algorithm 1 Membership function configuration

Require: Mp_k, Li_k and Ri_k, $k = 1 \ldots Q$;
Ensure: min_k, max_k, a_k e b_k, $k = 1 \ldots 2 \times Q$;
Ensure: min, max, a and b;
  if $enable = 1$ **then**
    $Address \leftarrow 0$;
    for $k \leftarrow 1\ to\ Q$ **do**
      for $FP \leftarrow 1\ to\ 2$ **do**
        $Write \leftarrow 0$; $Address \leftarrow Address + 1$;
        if $k = 1$ **then**
          if $FP = 1$ **then**
            $min \leftarrow -\infty$; $max \leftarrow Mp_k$; $a \leftarrow 0$; $b \leftarrow 1$;
          else
            $min \leftarrow Mp_k$; $max \leftarrow Mp_k + Ri_k$;
            $a \leftarrow -1/Ri_k$; $b \leftarrow (Mp_k/Ri_k) + 1$;
          end if
        end if
        if $1 < k < Q$ **then**
          if $FP = 1$ **then**
            $min \leftarrow Mp_k - Li_k$; $max \leftarrow Mp_k$;
            $a \leftarrow 1/Li_k$; $b \leftarrow -((Mp_k - Li_k)/Li_k)$;
          else
            $min \leftarrow Mp_k$; $max \leftarrow Mp_k + Ri_k$;
            $a \leftarrow -1/Ri_k$; $b \leftarrow (Mp_k/Ri_k) + 1$;
          end if
        end if
        if $k = Q$ **then**
          if $FP = 1$ **then**
            $min \leftarrow Mp_k - Li_k$; $max \leftarrow Mp_k$;
            $a \leftarrow 1/Li_k$; $b \leftarrow -((Mp_k - Li_k)/Li_k)$;
          else
            $min \leftarrow Mp_k$; $max \leftarrow +\infty$; $a \leftarrow 0$; $b \leftarrow 1$;
          end if
        end if
        $write \leftarrow 1$;
      end for
    end for
  end if

The obtained results for the two straight lines modeling the linguistic term are kept in two distinct 32-bit registers. These are the truth degrees, once it is delivered by the FPU. The block includes two sets of 32-bit registers, namely TuFP1 and TuFP2, one for each linguistic term modeling the membership function of the input variable.

The inputs of a Fuzzy block are the characteristics of the linguistic terms of the membership function associated with the input variable under consideration. These characteristics are a, b, min and max stored in MF MEM segment corresponding to the input variable, as explained in Sect. 4.5.2. The output of a Fuzzy block are: signal EnF_i, for $i = 1 \ldots Q$ bits, i.e., one for each included linguistic term and signal uF_i,

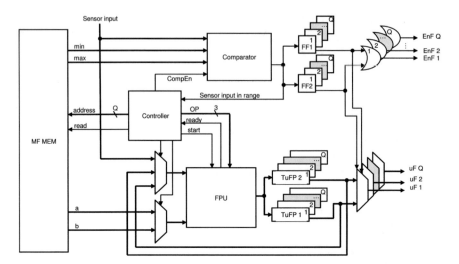

Fig. 4.6 Fuzzy block micro-architecture

for $i = 1 \ldots Q$ 32-bit floating-point values, each of which represents the truth degree of the corresponding linguistic term. Note that linguistic terms that do not apply have 0 as a truth degree. When bit EnF_i is activated, this indicates that linguistic term number i of the membership function is valid with truth degree $uF_i \neq 0$. Recall that the truth degree is the product of a and input value augmented by b. In Algorithm 2, we give an overview on how the Fuzzy block operates.

4.5.4 Inference Unit

The inference unit main purpose is to identify, for each one of the output variables of the fuzzy controller, the linguistic terms that are active as well as computing the associated truth degrees.

Before describing the details of the inference unit, let us first introduce the structure used to format the rules of the fuzzy system. A rule $\mathcal{R}$ has two defining parts: a premise $\mathcal{P}$ and a consequent $\mathcal{C}$ as described in (4.4), wherein $\mathcal{I}_i$, for $i = 1 \ldots N$ are the input variables and $T_k^{\mathcal{I}_i}$ for $k = 1 \ldots Q$ are the linguistic terms associated to it, $\mathcal{O}_j$, for $j = 1 \ldots M$ are the output variables and $T_k^{\mathcal{O}_j}$ for $\ell = 1 \ldots Q$ are the linguistic terms associated with it. Note that in general, the number of linguistic terms is distinct from one variable to another. However, in this work, we assume, without loss of generality, that all the variables, both of input and output, are modeled using the same number of linguistic terms Q. A rule may check only few of of the N input variables, and it may also, enable only few of the output variables.

Algorithm 2 Operation of the Fuzzification block

Require: $sensor, min, max, a$ e b;
Ensure: uF_i e EnF_i, $i = 1 \dots 2 \times Q$;
  if $enable = 1$ **then**
    for $i \leftarrow 1 \, to \, 2 \times Q$ **do**
      $Address \leftarrow i; read \leftarrow 1$;
      if $min < sensor < max$ **then**
        $FPUout_i \leftarrow sensor \times a + b; COMPout_i \leftarrow 1$;
      else
        $FPUout_i \leftarrow 0; COMPout_i \leftarrow 0$;
      end if
    end for
    $read \leftarrow 0; k \leftarrow 1$;
    for $i \leftarrow 1 \, to \, Q$ **do**
      $EnF_i \leftarrow (COMPout_k \; OR \; COMPout_{k+1})$;
      if $COMPout_k = 1$ **then**
        $uF_i \leftarrow FPUout_k$;
      else if $COMPout_{k+1} = 1$ **then**
        $uF_i \leftarrow FPUout_{k+1}$;
      else
        $uF_i \leftarrow 0$;
      end if
      $k \leftarrow k + 2$;
    end for
  end if

$$\mathcal{R}: \; \mathcal{P} \Rightarrow \mathcal{C}, \text{ where for } j, k, \ell = 0 \dots Q:$$
$$\mathcal{P} \text{ is } \mathcal{I}_0 = T_j^{\mathcal{I}_0} \wedge \mathcal{I}_1 = T_k^{\mathcal{I}_1} \wedge \cdots \wedge \mathcal{I}_{N-1} = T_\ell^{\mathcal{I}_{N-1}} \qquad (4.4)$$
$$\mathcal{C} \text{ is } \mathcal{O}_0 = T_j^{\mathcal{O}_0} \wedge \mathcal{O}_1 = T_k^{\mathcal{O}_1} \wedge \cdots \wedge \mathcal{O}_{N-1} = T_\ell^{\mathcal{O}_{N-1}}$$

The rule base memory `Rules` has a word size that allows to store one rule. All the rules of the model have the same structure. They include all the input and output variables. When a variable is not checked or inferred, the all the linguistic terms are checked off.

A given rule fires when signal EnF_i, as delivered by the FU, for every checked of linguistic term of every input variable of the premise part of the rule under consideration is set. Furthermore, every linguistic term of any output variable that is checked in the consequent part of a fired rule need to be reported to the defuzzification unit FU. Note that there are at most M, one for each output variable. Besides this, FU needs also to receive the truth degree for each of these checked terms.

The truth degree of an output variable linguistic term is the smallest truth degree, considering all those associated with the input variable linguistic terms in the premise part of the fired rule. When the same output variable linguistic term appears on two or more fired rules, the highest truth degree is used. Thus, this done considering all the rules that fires. Recall that the truth degree of the input variable linguistic terms are provided by the FU.

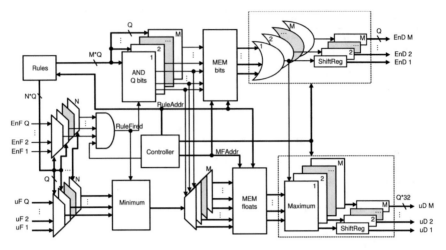

Fig. 4.7 Inference block micro-architecture

Figure 4.7 shows the micro-architecture of the `Inference` block. Its inputs consist of the Q flags EnF_i, for $i = 1 \ldots Q$ and the corresponding Q truth degrees uF_i, for $i = 1 \ldots Q$, which are the resulting output of FU, as described in Sect. 4.5.3. Its outputs are a set of M Q-bit signals EnD_i, for $i = 1 \ldots M$, that identify the linguistic terms that were inferred and their respective truth degrees uD_i, for $i = 1 \ldots M$, which are signals of $Q \times 32$ bits. The AND gate determines wither the current rule can be fired. In Algorithm 3, we sketch how the operation of the inference block is controlled. The M `ANDQbits` components are simply na AND-arrays. In this design, the process of min-max inference is used. So, components `Minimum` and `Maximum` return the smallest of N floats and the highest of M floats, respectively. Their internal structure is omitted here for a lack of space. The `Inference` includes three memory blocks: the rule base `Rules`, a truth degree memory `MEM floats` and a bit memory `MEM bits`.

4.5.5 Defuzzification Unit

The defuzzification unit's main purpose is to compute the crisp value of the output variables, given the fuzzy linguistic terms and their corresponding truth values, as identified and computed by the inference unit. The centroid is used to perform the defuzzification process. Recall that uD_i for $i = 1 \ldots Q$ are the truth degrees of the linguistic terms associated with the output variable $\mathcal{O}$. The computation is done according to the steps of Algorithm 4.

Algorithm 3 Inference control and computation

Require: uF_i^j, EnF_i^j, $i = 1 \ldots Q$, $j = 1 \ldots N$ and $Rules_r^l$, $r = 1 \ldots P$, $l = 1 \ldots (N + M) \times Q$;

Ensure: uD_i^k e EnD_i^k, $k = 1 \ldots M$, $i = 1 \ldots Q$;

  if $enable = 1$ **then**
    for $r \leftarrow 1\ to\ P$ **do**
      $\mathcal{R} \leftarrow Rules_r$;
      if $\mathcal{R}$ Valid **then**
        for $j \leftarrow 1\ to\ N$ **do**
          $\mathcal{I} \leftarrow \mathcal{R}^j$; `AndInput`$_j \leftarrow \mathcal{I}$ & EnF^j; `MinInput`$_j \leftarrow uF^j$;
        end for
        $RuleFired \leftarrow$ AND($AndInput$);
        if $RuleFired$ **then**
          $Min \leftarrow$ MIN($MinInput$);
          for $k \leftarrow 1\ to\ M$ **do**
            $\mathcal{O} \leftarrow \mathcal{R}^{N+k}$;
            for $i \leftarrow 1\ to\ Q$ **do**
              if $\mathcal{O}_i = 1$ **then**
                `MEMfloats`$_r^{k \times i} \leftarrow Min$; `MEMbits`$_r^k \leftarrow \mathcal{O}$;
              else
                `MEMfloats`$_r^{k \times i} \leftarrow 0$; `MEMbits`$_r^k \leftarrow 0$;
              end if
            end for
          end for
        end if
      else
        `MEMfloats`$_r \leftarrow 0$; `MEMbits`$_r \leftarrow 0$;
      end if
    end for
    for $k \leftarrow 1\ to\ M$ **do**
      for $i \leftarrow 1\ to\ Q$ **do**
        $uD_i^k \leftarrow$ MAX(`MEMfloats`$^{k \times i}$); $EnD_i^k \leftarrow$ OR(`MEMbits`$^{k \times i}$);
      end for
    end for
  end if

Algorithm 4 Computation of the centroid

  $R_0 \Leftarrow 0$; $R_1 \Leftarrow 0$; $R_2 \Leftarrow 0$;
  if $EnD \neq 00 \ldots 0$ **then**
    for $i := 1\ to\ Q$ **do**
      if $EnD_i = 1$ **then**
        $R_0 \Leftarrow uD_i \times mp_i$;
        $R_1 \Leftarrow R_1 + R_0$; $R_2 \Leftarrow R_2 + uD_i$;
      end if
    end for
    $R_0 \Leftarrow R_1/R_2$;
  end if **return** R_0;

4.6 Performance Results

The application presented in the following is for the angular velocity control. It requires two input variables that shape the radius and angle in polar form representing the error and error variation of speed, 15 fuzzy rules as described in Table 4.1, 5 linguistic terms and 1 output variable that represents the linear velocity of the robot movement. Figure 4.8 shows the membership functions used for each of the input and output variables.

Table 4.2 shows the sensor input values tested, the rules fired, according to Table 4.1. Also it shows the number of linguistic terms that were activated at the start of the defuzzification process, the number of clock cycles, the execution time in microseconds, based on clock of 112.410 MHz and the scalar value of the result of the hardware controller.

Figure 4.9 shows the control surface based on the configuration of the fuzzy controller for this application. The computation of the quadratic error, as defined in (4.5), is 3.1237×10^{-7}, which shows an excellent accuracy in comparison to the software implementation using MATLAB. In (4.5), xh_i is the ith result returned by the recon-

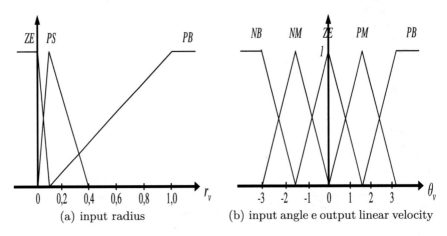

(a) input radius   (b) input angle e output linear velocity

Fig. 4.8 Membership function used

Table 4.1 Fuzzy rules for the autonomous robot navigation

Rule		Radius		
		ZE	PS	PB
Angle	PB	r_0: ZE	r_1: NM	r_2: NB
	PM	r_3: ZE	r_4: PM	r_5: PB
	ZE	r_6:ZE	r_7: PM	r_8: PB
	NM	r_9: ZE	r_{10}: NM	r_{11}: NB
	NB	r_{12}:ZE	r_{13}: NM	r_{14}: NB

Table 4.2 The results obtained by the the reconfigurable hardware for the robot navigation control

Radius	Angle	Rules fired defuzzy	Number of defuzzy	Cycles clock	Time (μseg)	Velocity
0.00	+1.0	r_3 e r_6	1	1043	9.28	0.0000
0.80	−2.0	r_{11} e r_{14}	1	1043	9.28	−3.0000
0.50	+2.5	r_2 e r_5	2	1090	9.70	−0.4286
0.05	−1.0	r_6, r_7, r_9 e r_{10}	3	1137	10.11	−0.1875
0.09	+2.0	r_0, r_1, r_3 e r_4	3	1137	10.11	+0.4545
0.30	+2.0	r_1, r_2, r_4 e r_5	4	1184	10.53	0.0000
0.20	−0.5	r_7, r_9, r_{11} e r_{12}	4	1184	10.53	+0.4091
0.20	+2.5	r_1, r_2, r_4 e r_5	4	1184	10.53	−0.4091

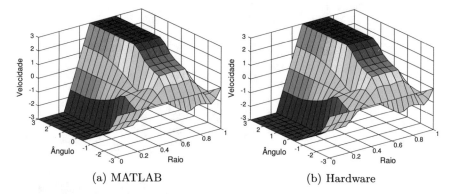

(a) MATLAB   (b) Hardware

Fig. 4.9 Control surface for the wheeled robot navigation

figurable controller hardware, xm_i is the ith result of returned by the toolbox FIS of MATLAB, and n represents the total number of obtained results. In this case, we use 17 distinct set of inputs.

$$Error = \frac{\sum_{i=1}^{n}(xh_i - xm_i)^2}{n} \tag{4.5}$$

Using a clock frequency of 100 MHz in FPGA, the entire controller runs, in the worst case, with Defuzzy4 in 2,246 clock cycles, i.e., 22.46 μs. The synthesis results show that the maximum clock frequency accepted by the design developed for this application is 112.410 MHz, which resulted in an execution time of 19.98 μs. As the operation time of block FP is not accounted for in the normal cycle of the controller loop, the latter will be at most of of 1,184 clock cycles, i.e., 10.53 μs, considering the maximum allowable clock frequency. Figure 4.10 displays the number of clock cycles for each block of reconfigurable controller, including variations in numbers of cycles for block Fuzzy. Figure 4.11 shows the execution times of each block, using the maximum clock frequency.

Fig. 4.10 Number of clock cycles required by the reconfigurable controller

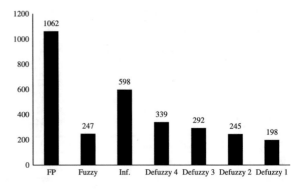

Fig. 4.11 Execution time, in microseconds, of the blocks in the FPGA

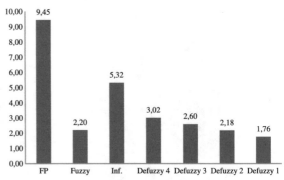

Fig. 4.12 Hardware area usage in the FPGA

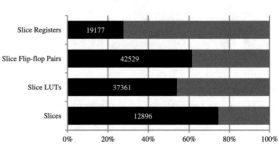

Figure 4.12 shows the hardware area required in the FPGA to program the entire fuzzy controller. Considering the 69,120 LuTs available in the FPGA, only 54.1 % and used.

4.7 Conclusion

This paper proposes a massively parallel completely configurable design for fuzzy controllers. It is applicable to almost any applications in the industry that do not have a prescribed solution. The proposed architecture is parametric so that any number of inputs, outputs, and rules can be accommodate with no extra effort. The design was implemented on reconfigurable FPGA and the cost and performance requirements analyzed. The fuzzy controller supervises the navigational process of a wheeled robot. The next steps in the design of this controller are to investigate the generalization of the design so that to allow the use of trapezoidal and sigmoid the membership functions.

References

1. Lin WS, Huang CL, Chuang MK (2005) Hierarchical fuzzy control for autonomous navigation of wheeled robots. IEE Proc-Control Theory Appl 152(5):598–606
2. McKenna M, Wilamowski B (2001) Implementing a fuzzy system on a field programmable gate array fuzzy sets and systems. In: Proceedings IJCNN '01 - international joint conference on neural networks, vol 1, pp 189–194. IEEE, Washington, DC
3. Pappis CP, Mamdani EH (1977) A fuzzy logic controller for a traffic junction. IEEE Trans Man Cybern 7(10):707–717
4. Poorani S, Urmila Priya T, Udaya K, Renganarayanan S (2005) FPGA based fuzzy logic controllers for electric vehicle. J Inst Eng 45(5):1–14
5. Zadeh L (1968) Fuzzy algorithms. J Inf Control 12(2):94–102
6. Zadeh L (1984) Making computers think like people. IEEE Spectr 21(8):26–32
7. Zadeh L (1988) Fuzzy logic. IEEE Comput J 21(4):83–93

Chapter 5
Nonlinear Correction for an Energy Estimator Operating at Severe Pile-Up Conditions

**Bernardo Sotto-Maior Peralva, Alessa Monay e Silva,
Luciano Manhães de Andrade Filho, Augusto Santiago Cerqueira
and José Manoel de Seixas**

For systems operating at high event rates, the readout signal may be distorted by the presence of information from adjacent events. The signal superposition, or pile-up, degrades the efficiency of linear methods, which are typically used for signal parameter estimation. In many applications, the estimation task reduces to determine the amplitude of the incoming signal. In the context of high-energy calorimeters, which aim at measuring the energy of high-energy subproducts of interactions, the signal energy is measured by estimating the amplitude of the received digitized pulse. Modern particle colliders may operate at an event rate much higher than their calorimeter time response length and, as a result, the signal pile-up may be observed. This chapter describes how a computational intelligence approach can assist on the energy estimation performed by an optimal linear method. An artificial neural network is trained aiming at correcting for the nonlinearities introduced by the signal pile-up statistics. The efficiency of the various energy estimation methods is evaluated from simulation data under various signal pile-up scenarios.

B.S.-M. Peralva (✉) · A.M. e Silva · L.M. de Andrade Filho · A.S. Cerqueira
Electrical Engineering Graduate Program, Federal University of Juiz de Fora,
Juiz de Fora, Brazil
e-mail: bernardo@cern.ch

A.M. e Silva
e-mail: alessa.monay@engenharia.ufjf.br

L.M. de Andrade Filho
e-mail: luciano.andrade@ufjf.edu.br

A.S. Cerqueira
e-mail: augusto.santiago@ufjf.edu.br

J.M. de Seixas
Signal Processing Laboratory, Federal University of Rio de Janeiro, Rio de Janeiro, Brazil
e-mail: seixas@lps.ufrj.br

© Springer International Publishing Switzerland 2017
N. Nedjah et al. (eds.), *Designing with Computational Intelligence*,
Studies in Computational Intelligence 664,
DOI 10.1007/978-3-319-44735-3_5

5.1 Introduction

Parameter estimation has been used in several areas such as communication, economy and finance, instrumentation and experimental physics [8]. A signal, usually corrupted from noise, must carry information that is usually related to one or a set of signal parameters such as amplitude, phase, frequency, length, etc. Basically, the parameter estimation task consists in accurately compute a physical quantity, envisaging the extraction of the desired information from the signal. Particularly, the problem of estimating the amplitude of a signal appears in various aspects in signal processing [25]. Due to their simplicity, efficiency and fast response, linear filters are extensively employed for parameter estimation tasks in online applications [34, 40, 48]. However, under hash conditions, where additional nonlinear components are strongly present in the noise, the performance of linear methods tends to decrease, introducing bias and increasing uncertainties in the measure. The solution often leads to nonlinear parameter estimation [7, 35, 51].

Computational intelligence techniques are widely applied in the context of parameter estimation in areas such as computer vision, biometrics, communication, among others [26, 27, 29]. In experimental high-energy physics (HEP), the use of computational intelligence is growing due to the increased complexity of the modern experiments along with the advances in hardware processing capabilities [24, 41, 49]. Among the most common computational intelligence techniques used in HEP, Artificial Neural Networks (ANN) [22] can be found in numerous applications such as pattern recognition, track reconstruction, triggering and physics analysis [2, 14, 42, 46, 47], as well as for energy estimation tasks [20, 38, 39].

Particle colliders are complex facilities used in experimental HEP [28]. They are built to accelerate particles and perform collisions at very high energy levels (order of Tera-electron Volts). Essentially, these experiments cover a large physics research program, and the physicists are interested in explaining some of the fundamental building blocks that have shaped the universe. To this end, two beams (or one beam) of particles traveling at approximately the speed of light are put head-to-head (or made collide with a fix target) and the subproduct of these collisions are measured by high precision detectors. Several scientific discoveries were made possible through the use of particle accelerators including, more recently, the existence of the Higgs boson, a particle that is foreseen in the Standard Model [13] but never observed until 2015 [45].

Since the physics of interest is rare, a massive quantity of data is needed in order to infer any behavior in the data. Therefore, modern colliders tend to increase their beam luminosity so that more signals become available and the probability of observing a rare process increases. The beam luminosity is proportional to the number of interactions per second divided by the beam cross section area [23]. With the increase of the luminosity, the beam becomes denser, so more interactions may occur at each beam (or bunch) crossing [36].

High-energy calorimeters play an important hole as they absorb and sample the energy of the incoming particles, providing precise measurements of the energy flow and also being used for triggering and event selection tasks [33]. Particularly, in high-event rate experiments, such as the Large Hadron Collider (LHC) [16], the calorimeter response is usually slower than the event rate, so that the readout window may comprise several bunch crossings. As a result, the effect of signal superposition (also called signal pile-up) may be observed within the readout window, distorting the signal of interest and degrading the final energy estimation [17, 31].

The classical methods used in calorimetry for online amplitude estimation are based on linear filtering, whose coefficients are found by an optimization procedure that minimizes the estimator variance [6]. They are simple, fast and operate in optimal conditions when the calorimeter pulse shape must be fixed and corrupted from additive Gaussian noise only. In cases where the pulse shape presents fluctuations and/or the additive noise is not Gaussian (due to the presence of signal pile-up, for instance), the efficiency of such methods is degraded.

An alternative technique uses a computational intelligence approach to estimate the energy under conditions of high luminosity. In order to handle the nonlinearities produced by signal pile-up, a ANN may be designed either to perform the full amplitude (energy) estimation or to assist the linear energy estimation. In the case of assisting the linear estimation, the energy is estimated by an optimal linear method based on Maximum Likelihood Estimation (MLE) [25], and a feedforward Multi-Layer Perceptron (MLP) [22] is trained to compensate for the error introduced by the nonlinear component that is present in the noise due to the signal pile-up. In this way, the ANN performs a fine adjustment to the linear estimate in order to correct for it. The linear estimate (MLE) remains available and preserved in case the nonlinear correction is needless.

The use of a ANN for nonlinear correction has been applied in other context as well [30, 50]. Furthermore, the use of a simple (low computational complexity) MLP for online amplitude estimation in calorimeters are particularly interesting, since one of the requirements for this task is of the low computational cost as the estimation is to be performed at high rates (tens of MHz) and for hundred of thousand channels [12, 43, 44]. With the growing capacity of devices such as the ones from FPGA (Field Programmable Gate Arrays) [32] technology, the use of neural networks for online energy estimation has become feasible.

The text is organized as follows. In the next section, the high-energy calorimetry environment is briefly introduced. In Sect. 5.3, the most commonly used algorithms for online energy estimation are described. Section 5.4 presents, in details, the use of a ANN for energy estimation for a high rate general-purpose calorimeter system. The simulation results, considering different signal pile-up conditions, are shown in Sect. 5.5. Finally, the conclusions are outlined in Sect. 5.6.

5.2 High-Energy Signal Pile-Up

Calorimetry systems play an important role in high-energy colliders. The information provided by the calorimeters is the key element to understand the physics processes. Sampling calorimetry is the most used technology. These high-energy detectors are in charge of absorbing, sampling, and performing precise measurements of the energy of the incident particles. Typically, they are finely segmented and present thousands of readout channels. The particles that are produced from the collisions cross and interact with the calorimeter, and their energy is sampled and converted to an electrical signal that is fed into a readout electronics.

The analog signal is conditioned in such a way that its shape remains almost invariant from channel to channel and the amplitude proportional to the signal energy [3]. Therefore, by estimating the signal amplitude, the energy can be found. The shaped signals from readout channels are typically digitized at the same rate as the collision rate.

The signal samples are sent to digital processors where the digital processing, which computes the signal amplitude, is finally carried out. Figure 5.1a shows a typical asymmetric unipolar pulse (with approximately 150 ns of width and sampled at 100 MHz) used in modern calorimeter systems as well as its digital samples.

Fig. 5.1 **a** Typical asymmetric unipolar calorimeter pulse sampled at 100 MHz and **b** pile-up signal at t = 50 ns

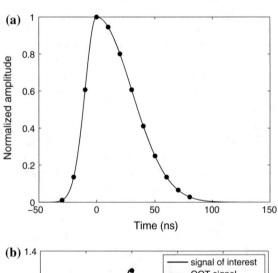

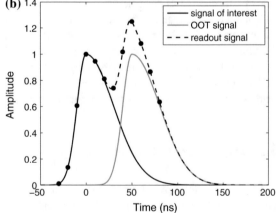

The effect of the signal pile-up in this type of readout pulse is seen in Fig. 5.1b, where the signal of interest is distorted by the presence of an Out-Of-Time (OOT) signal peaking at 50 ns and within the same readout window of 150 ns, resulting in a deformed readout signal.

Typically, the energy estimation in modern calorimetry is carried out through a linear combination of the received digital signal [6]. The noise covariance matrix is usually used in the optimization process that computes the coefficients in order to reduce the uncertainties in the measurements due to the electronic noise of each readout channel. However, the information from the signal pile-up introduces non-linearities in the background so that the noise distribution is no longer Gaussian. Therefore, the efficiency of linear techniques is degraded and they become sub-optimal. Recently, a new technique based on linear signal deconvolution has been proposed [17]. However, it addresses offline energy reconstruction and it is designed for signals that remain within the readout window.

5.3 Online Linear Estimation

In this section, the commonly used algorithms for online energy estimation in calorimetry are described. The first one is the most employed algorithm in the area, while the second one corresponds to an alternative design approach that may lead to a similar estimator, depending on the assumptions that are made regarding the signals involved.

5.3.1 Minimum Variance Linear Unbiased Estimator

In most modern calorimeters, noise comprises mainly electronic noise that is often assumed stationary and modeled by a Gaussian distribution. Therefore, variance minimization techniques for energy reconstruction are extensively employed as they perform close to the optimal operation. It should be stressed that the parameters of instability of the pulse, such as deformation, also introduce uncertainties to the final energy estimation, and they are not taken into account in the design of typical energy estimators.

Most algorithms for energy reconstruction in calorimeter systems are based on a weighted sum [6]. The signal amplitude $\hat{A}$ is estimated through a linear combination of the discrete time window that contains the readout signal $s[k]$ of N samples, where k corresponds to the time samples:

$$\hat{A} = \sum_{k=0}^{N-1} w[k]s[k] \tag{5.1}$$

The weights $w[k]$ are obtained from the front-end pulse shape and the noise covariance matrix. The procedure aims at minimizing the variance of the amplitude distribution. Thus, they are optimal for deterministic signals corrupted by Gaussian noise. The correct weights are computed by minimizing the effect of the noise in the amplitude reconstruction.

In order to compute the weights $w[k]$, the calorimeter signal can be modeled through a first-order approximation, aiming at linearizing the pulse phase parameter:

$$s[k] = Ag[k] - A\tau g'[k] + n[k] + ped \qquad k = 0, 1, 2, \ldots, N-1 \qquad (5.2)$$

where $s[k]$ represents the received digital time sample k and N is the total number of samples available [18]. The amplitude A is the parameter to be estimated, while $n[k]$ is the background noise. The parameters $g[k]$ and $g'[k]$ are the reference pulse shape and its derivative, respectively, while the parameter τ is the signal phase. The variable ped corresponds to the signal pedestal and it is a constant value added to the analog signal just before its analog-to-digital conversion.

For an unbiased estimator, it is required that the expected value of $\hat{A}$ to be A. Therefore, Eq. (5.3) can be derived for an optimal filter, when the noise is assumed as a zero-mean random process $(E[n[k]] = 0)$.

$$E[\hat{A}] = \sum_{k=0}^{N-1} (Aw[k]g[k] - A\tau w[k]g'[k] + w[k]ped) \qquad (5.3)$$

The following constraints can be deduced in order to reach $E[\hat{A}] = A$ [18]:

$$\sum_{k=0}^{N-1} w[k]g[k] = 1 \qquad (5.4)$$

$$\sum_{k=0}^{N-1} w[k]g'[k] = 0 \qquad (5.5)$$

$$\sum_{k=0}^{N-1} w[k] = 0 \qquad (5.6)$$

These constraints are added to the minimization procedure in order to reduce the uncertainties due to pedestal fluctuations and phase shifts. On the other hand, the imposition of these constraints may increase the variance of the estimator. Thus, for example, in cases where the electronic pedestal value can be accurately measured and subtracted from the incoming digitized signals, the constraint expressed in Eq. (5.6) can be removed, increasing the number of degrees of freedom of the optimization procedure [11].

The variance of the estimator is given by,

$$var(\hat{A}) = \sum_{k=0}^{N-1}\sum_{j=0}^{N-1} w[k]w[j]C[k,j] = \mathbf{w}^T\mathbf{C}\mathbf{w} \qquad (5.7)$$

where $\mathbf{C}$ corresponds to the noise covariance matrix and $\mathbf{w}$ is the vector of weights. Hence, to find the optimal weights, Eq. (5.7) is often minimized subjected to the constraints in (5.4), (5.5) and maybe (5.6) using Lagrange multipliers [6].

Most of the modern calorimeters uses the described algorithm to reconstruct the signal amplitude [1, 6, 11]. However, each algorithm is slightly modified from one system to another to estimate the final amplitude, particularly in the way the signal baseline value is used.

5.3.2 Maximum Likelihood Estimation

In the Maximum Likelihood Estimator (MLE) [25] approach, the estimation problem is formulated from the probability density functions of the random process. The first step in the MLE design is to compute the probability density functions of the received signal $p(\mathbf{s}|\hat{A}_{mle})$, given that it has an amplitude A to be estimated and $\hat{A}_{mle}$ corresponds to the MLE amplitude estimate. The best estimate of A is the value that maximizes $p(\mathbf{s}|\hat{A}_{mle})$. Therefore, the amplitude estimate can be found by solving the following equation for the $\hat{A}_{mle}$:

$$\frac{\partial p(\mathbf{s}|\hat{A}_{mle})}{\partial A} = 0 \qquad (5.8)$$

The a priori knowledge of the random process described by the signal is necessary. For the sake of simplicity, initially, the phase of the received pulse may be considered fixed for each readout channel. Additionally, the baseline value may be subtracted from each received digital sample before estimation. As a result, the input signal $\mathbf{s}$ considered in the MLE design becomes:

$$\mathbf{s} = A\mathbf{g} + \mathbf{n} \qquad (5.9)$$

where A represents the amplitude of the received signal, the vector $\mathbf{g}$ corresponds to the samples of the normalized reference pulse and $\mathbf{n}$ are the noise samples.

For the particular case where the noise samples can be modeled by a multivariate Gaussian distribution with covariance matrix $\mathbf{C}$, the probability density function is given by the following expression:

$$p(\mathbf{s}|\hat{A}_{mle}) = \frac{1}{\sqrt{2\pi \det(\mathbf{C})}} \exp\left(\frac{-(\mathbf{s} - A\mathbf{g})^T\mathbf{C}^{-1}(\mathbf{s} - A\mathbf{g})}{2}\right). \qquad (5.10)$$

By extracting the logarithm in (5.10), we have:

$$\log\{p(\mathbf{s}|\hat{A}_{mle})\} = \frac{-1}{\sqrt{2\pi \det(\mathbf{C})}} \frac{(\mathbf{s} - A\mathbf{g})^T \mathbf{C}^{-1}(\mathbf{s} - A\mathbf{g})}{2}. \tag{5.11}$$

Computing the derivative of (5.11), with respect to the amplitude, and setting the result to zero, it leads to:

$$\hat{A}_{mle} = \frac{\mathbf{s}^T \mathbf{C}^{-1}\mathbf{g}}{\mathbf{g}^T \mathbf{C}^{-1}\mathbf{g}} = \mathbf{s}^T \mathbf{w} = \sum_{k=0}^{N-1} s[k]w[k], \tag{5.12}$$

where $\hat{A}_{mle}$ is the amplitude estimate that maximizes Eq. (5.10).

Similarly to other linear methods, for the considered constraints, the MLE can be easily implemented in digital processors through a Finite Impulse Response (FIR) filter. It should be pointed out that if the constraint expressed by Eq. (5.6) is suppressed, the coefficients of the linear estimator described in Sect. 5.3.1 lead to the $\mathbf{w}$ vector computed for the MLE in Eq. (5.12) [25].

The Gaussian noise hypothesis is valid for the case where the signal pile-up effect is not likely to occur. However, as the MLE uses the joint probability density function, the pile-up effect could be incorporated to its design. On the other hand, the use of the correct signal pile-up model requires an a priori knowledge, which is not available in the majority of the applications. Besides, it may result in a complex estimator that does not have analytical solution or be difficult to implement.

5.4 Nonlinear Correction

The linear methods do not fully handle the nonlinearities introduced in the received signal by signal pile-up. Therefore, a computational intelligence approach could be used in order to assist the reconstruction of the energy performed by typical linear techniques. A nonlinear corrector may be designed in order to provide a small contribution to the linear estimate according to the pile-up condition. For this, the MLE can be combined with a ANN.

Instead of using the actual pile-up noise statistics in the design of a non-Gaussian MLE, a combined model uses a standard MLE (designed for a Gaussian noise approximation) and a nonlinear corrector, which may be implemented through an ANN (Fig. 5.2).

The goal of the nonlinear processing is to correct for the linear model. That is, the nonlinear corrector does not estimate the signal energy itself, but it provides an adjustment to the linear estimate already available. For the case where the noise comes mainly from electronic noise, the nonlinear contribution should be minimal, and the estimate is dominated by the linear method.

Fig. 5.2 Block diagram illustrating the combined system to perform the energy estimation under severe signal pile-up conditions. A linear estimator is combined with a nonlinear corrector that performs an adjustment for the final estimate

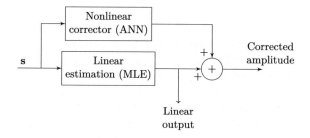

The advantage of the combined system is that the estimation from the linear estimator (MLE) is always available for use, and the correction is applied upon the user decision. Furthermore, the nonlinear corrector aims at adjusting the energy that has already been computed. Therefore, its design should be simpler in terms of computational effort, which is attractive for online operation.

5.5 Performance Comparisons

For performance comparisons, general-purpose simulation data are used here. The 12-sample pulse shown in Fig. 5.1a was considered as the reference pulse shape coming from the calorimeter electronics front-end. Additionally, the simulation considers a collider operating at the same rate as the front-end ADC sampling frequency. It is worth mentioning that the simulations use 64-bit floating-point numbers and no quantization issues were considered.

In order to cover the main algorithms used in modern calorimeter system, two different algorithms (linear filters) are considered. The first one, referred to here as Method 1, does not use the pedestal constraint (Eq. (5.6) removed), but, instead, uses a baseline value stored in a data base. The second one, called Method 2, applies the pedestal constraint in its design. Additionally, a feedforward MLP designed to provide the complete estimate of the energy is also used for comparison with the nonlinear correction approach. The ANN designed to perform the full estimation replaces entirely the linear model aiming at approximating the optimum estimator [37]. This will be referred to as ANNE (Artificial Neural Network Estimator).

5.5.1 Data Set

A data set was built considering only signals corrupted with a zero-mean Gaussian noise ($\sigma = 1$ ADC count), which is typical in modern calorimeters [5, 15, 19]. The amplitude of the signals was chosen randomly according to an exponential distribution having a mean value set to 300 ADC counts. In order to simulate real operation conditions, another two random parameters were considered when generating the signals. An uniformly distributed variable ranging from 2 % of the sampling rate period simulates the phase shift due to the time-of-flight of the particle. Another zero-mean Gaussian distributed variable ($\sigma = 1$ %) emulates the pulse deformation due to electronics aging and precision.

For the first scenario of study, the signal of interest will not be superimposed with any other signal within the same readout window, i.e. no signal pile-up is taken into account (0 % of channel occupancy). Here, occupancy means the probability of a collision to produce a valid signal that is read out by a given channel. In a high signal occupancy channel, the signal of interest is likely to be corrupted with both electronic noise and signal pile-up.

Figure 5.3a shows the histogram of the signal samples along with a Gaussian fit. The kurtosis for the noise distribution in the scenario of 0 % of occupancy (no pile-up) is 3.02 (a pure Gaussian distribution has kurtosis equal to 3.00) and the hypothesis

Fig. 5.3 Distribution of the noise samples for the cases of **a** 0 % and **b** 50 % of occupancy

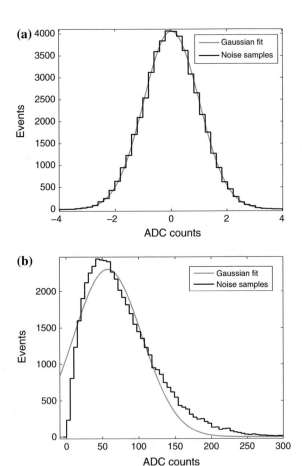

that the noise is a normal distribution, measured from the $\chi 2$ test, can not be rejected at a 5 % level of significance.

Another data set was built aiming at evaluating the efficiency of the methods under different conditions of signal pile-up. The signal pile-up is added on the top of the electronic noise and it is modeled by an exponential distribution whose mean is set to 30 ADC counts [4, 10]. This data set characterizes the readout channels that suffer from the signal pile-up effect, either because they are close to the collision beam spot, or are located in a region with high signal occupancy. Again, fluctuations due to phase shift (2 % of the original sampling frequency) and signal deformation (1 %) are taken into account in the signal generation. The considered occupancy is 50 %.

Figure 5.3b shows the histogram of signal samples as well as a Gaussian fit for such 50 % of occupancy data. Now, the kurtosis is 4.76 and the $\chi 2$ test rejects the hypothesis that the noise is a random sample from a normal distribution at 5 % of significance level. It can be clear noticed a long positive tail due to the signal pile-up. Thus, the Gaussian features are much less present when compared to non pile-up data.

Each data set (0 and 50 % of occupancy) contains 50,000 events, where half of them (training set) were used to develop the linear methods and train the nonlinear corrector. The second half was used for ANN overtraining control and validation (test data set) of the considered algorithms.

5.5.2 Nonlinear Corrector Design

The nonlinear corrector based on ANN (Fig. 5.2) was implemented through a feed-forward MLP, chosen due to its simplicity and successful use in similar applications [30, 50]. The signal samples from the training data set were normalized so that the values of the samples range from 0 to 1. It was used a single hidden layer (five neurons) and a single neuron in the output layer, as it is shown in Fig. 5.4. The input nodes are the normalized incoming digital samples.

The activation function used for the neurons in the hidden layer was the hyperbolic tangent, whereas for the output neuron, a linear function is employed. The training algorithm was the Levenberg–Marquardt [21], given its efficiency in converging to the minimal of Mean Squared Error (MSE) function.

The target vector for the training phase corresponds to the absolute difference between the reference value (known from the simulation) and the MLE estimate, as shown in Fig. 5.5.

The number of neurons in the hidden layer was chosen based on the dispersion (RMS) of the estimation error distribution when applied to the test data set. Table 5.1 shows the behavior of the estimation error when the number of neurons in the hidden layer is varied for both cases of occupancy (0 and 50 %). It can be noticed tat the configuration with five neurons presents good tradeoff between complexity and efficiency (estimation error).

Fig. 5.4 Artificial neural
network topology used as a
nonlinear corrector

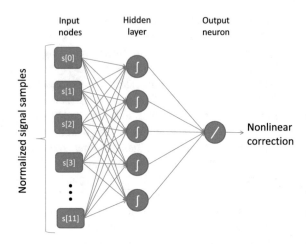

Fig. 5.5 ANN training
strategy used for the
combined method

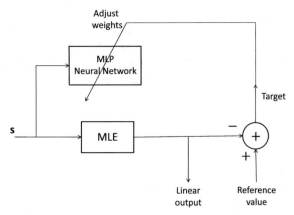

Table 5.1 Estimation error (RMS) for different numbers of neurons on the hidden layer (in ADC counts)

Occupancy	Number of neurons				
	2	3	4	5	10
0 %	0.9	0.9	0.9	0.9	0.9
50 %	28.5	28.2	27.6	27.4	27.4

5.5.3 Efficiency Tests

In order to evaluate the efficiency of the considered methods, the estimation error, implementation complexity and linearity are analyzed. The estimation error corresponds to the absolute difference between the estimated value and the reference value (from simulation). Figure 5.6 shows the estimation error distributions using the 0 % of occupancy data set, and the ANN output. As expected, the contribution from the

Fig. 5.6 Efficiency for the 0 % of occupancy test data: **a** estimation error for MLE + ANN (Mean = 0.0 and RMS = 0.9), Method 1 (Mean = 0.0 and RMS = 0.9) and Method 2 (Mean = 0.0 and RMS = 1.3); **b** ANN output (Mean = 0.0 and RMS = 0.2)

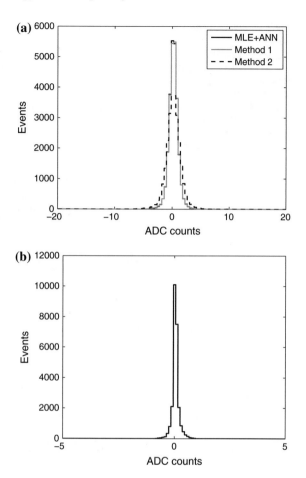

ANN to the final energy estimate is minimal (almost zero), as the nonlinearities due to the signal pile-up are absent for this scenario of occupancy.

Concerning the data set for 50 % of occupancy, Fig. 5.7 shows the estimation error and the contribution from the nonlinear correction (ANN output). As the signal pile-up introduces an additional energy to the signal of interest, as in Fig. 5.3b, the goal of the ANN is to compensate for this effect. For the sake of comparison, the estimation error for the MLE method without the ANN correction are 0.9 and 56.8 ADC counts for the case of 0 and 50 % of occupancy, respectively.

In both scenarios (occupancy of 0 and 50 %), the noise covariance matrix was used to design Methods 1 and 2 describing only the electronic noise (modeled by a multivariate Gaussian distribution). It should be mentioned that, Method 1 and MLE exhibit equivalent performance (same set of coefficients) and, although they present the optimum efficiency for the case of 0 % of occupancy, they do not take into account the signal pile-up statistics in their design. Therefore, they tend to present

Fig. 5.7 Efficiency for the 50% of occupancy test data: **a** estimation error for MLE + ANN (Mean = 0.2 and RMS = 27.4), Method 1 (Mean = 105.4 and RMS = 56.8) and Method 2 (Mean = −0.5 and RMS = 55.5); **b** ANN output (Mean = −105.2 and RMS = 49.8)

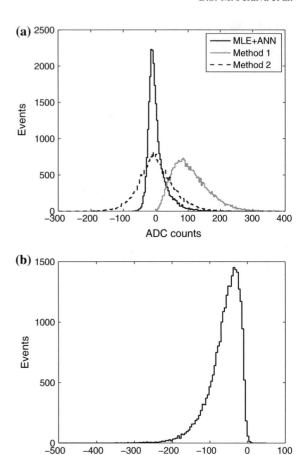

a positive bias as the superimposed signals introduce a positive offset energy to the distribution, which explains the model of the distribution from the nonlinear corrector output. As for Method 2, it presents an artifact (through an additional constraint in the optimization procedure) that forces the estimation to be immune against any offset, and its energy distribution tends to present a negative tail.

For the ANNE method, a MLP configuration with nine neurons is chosen (see Table 5.2). Figure 5.8 shows the estimation errors from the combined (MLE+ANN) and pure (ANNE) methods. It can be observed that the histograms are superimposed, which points out the similar performance of both approaches.

Concerning the computational effort, Table 5.3 shows the number of operations needed by each method. Method 1 and Method 2 require a low computational effort for operation, and they are suitable for online applications where resources are limited, as this is the case for most DSP (Digital Signal Processor) based systems.

Table 5.2   Estimation error (RMS) from the test set for different number of neurons in the hidden layer of the ANNE (in ADC counts), considering pile-up (50%) data

Occupancy	Number of neurons					
	3	7	8	9	10	20
50%	31.6	30.1	28.6	28.5	28.7	28.9

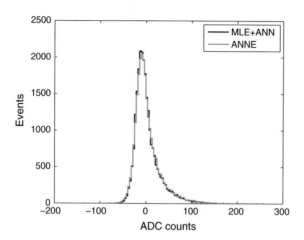

Fig. 5.8   Efficiency of the estimators based on ANN: estimation error for MLE + ANN (Mean = 0.2 and RMS = 27.4) and ANNE (Mean = O.2 and RMS = 28.5)

Table 5.3   Computational effort of different methods

	Operation		
	Sum	Product	Table search
Method 1	11	12	0
Method 2	11	12	0
MLE + ANN	60	65	5
ANNE	108	117	9

Considering nonlinear processing, the MLE+ANN approach presents better efficiency as it demands lower computational resources and the flexibility of having the linear output (MLE) available for use (upon user decision). It is worth mentioning that such combined method can be implemented and tested for online operation in experiments where modern electronic devices such as FPGA are employed [9].

Fig. 5.9 Linearity from the combined method for non pile-up data

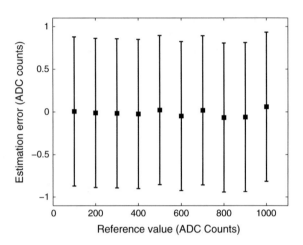

5.5.4 Bias and Linearity

The estimation error has to remain constant for a large operating dynamic range. Figure 5.9 shows the mean and the standard deviation of the absolute error using the combined method considering the scenario where the noise comes only from the electronic noise.

The data points and error bars remain both approximately constant along the energy range considered, indicating that the combined method presents a linear behavior and no bias is introduced in the final energy estimation.

5.5.5 Performance at Different Occupancy Levels

Let's consider now different levels of pile-up noise in order to cover the full range of channel occupancy. To this end, a data set for several levels of occupancy was built in a similar way described in Sect. 5.5.1 in order to cover a large range of occupancy. The case of 100 % of occupancy is the worst scenario of occupancy and it means that there is a valid signal in every bunch crossing. The ANN configuration and training strategy are kept the same as described in Sect. 5.5.2 for each nonlinear corrector associated to each occupancy level. Figure 5.10a shows the behavior of the mean and estimation error as a function of the occupancy for each of the methods. As expected, Method 1 exhibits a bias that increases with the occupancy. This bias can be parameterized and subtracted from each energy estimate according to the current beam luminosity (proportional to the occupancy).

Figure 5.10b shows the RMS value for the estimation error as a function of the occupancy level. It can be noted that for all occupancy levels, the MLE+ANN method presents better efficiency than Methods 1 and 2 showing that the nonlinearities are being properly dealt with.

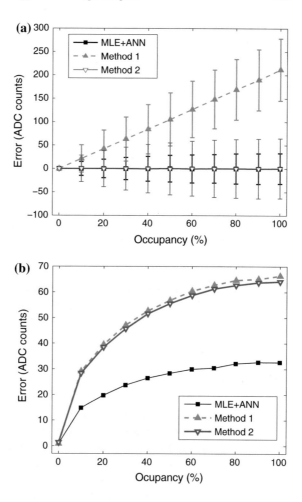

Fig. 5.10 Estimation error over different channel occupancy levels: the mean (**a**) and estimation error (RMS) (**b**)

5.5.6 Exploring the ANN Generalization

Because each collision reduces the density of the beam in real operation, typically the luminosity decreases as a function of time, and so does the channel occupancy. Therefore, an online application would require the set of ANN weights to be updated for every given luminosity during operation. However, a single ANN could be designed to generalize for all occupancy levels and the nonlinear corrector may become fully luminosity independent, avoiding reloading a specific set of ANN weights. Again, the ANN configuration and training strategy are kept the same as described in Sect. 5.5.2. Figure 5.11 shows the efficiency of such approach, called MLE+generalized ANN. A comparison with the ANN trained for each occupancy (MLE+ANN) is also provided.

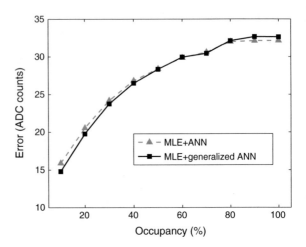

Fig. 5.11 Estimation error considering a large range of channel occupancy

The generalized ANN presents similar efficiency when compared to the ANN trained individually for each level of occupancy. For the case of 0 % of low channel occupancy (smaller than 10 %), the linear output may be indicated.

5.6 Conclusions

Due to simplicity, efficiency and fast response, linear filters have extensively been used for online energy estimation in modern high-energy calorimetry. These type of filters are suitable for calorimeters operating under low-luminosity, where the noise is wide-sense stationary and mainly comprises Gaussian noise arising from the electronic readout chain. They also meet implementation requirements when digital processing resources are limited. However, modern colliders tend to increase their luminosity, unavoidably leading to signal pile-up (higher channel occupancy). Since linear techniques do not incorporate the signal pile-up statistics in their design, it has been shown that they present large bias and low efficiency under pile-up conditions.

The noise corrupted by electronic noise plus signal pile-up is likely to present a non-Gaussian distribution due to the out-of-time signals that are acquired within a given readout window, resulting in a distribution with a larger positive tail. Besides, the signal pile-up statistics presents a non-stationary behavior since the level of luminosity often decreases with time during data taking. Therefore, linear methods based on a closed-form solution fail to properly describe the noise, indicating that the use of computational intelligence solutions may be appropriate for mapping the nonlinearities due to signal pile-up.

It was shown that an optimum estimator could be achieved by combining a linear estimator with a well-trained neural network, outperforming the usual linear techniques when operating at severe signal pile-up conditions. In this approach, the linear estimate is preserved and the ANN functions as a nonlinear corrector in order to compensate for the error introduced by the non-Gaussian components. Thus, the ANN may present low complexity, becoming attractive for online operation.

Finally, due to advances in hardware processing capabilities, the implementation of such sophisticated algorithms has become feasible in high-energy calorimetry, allowing intelligent systems to take part in future generations of particle colliders.

Acknowledgments The authors are thankful to CAPES, CNPq, RENAFAE (MCTI), FAPEMIG and FAPERJ (Brazil) for the support concerning this work.

References

1. Adzic P et al (2006) Reconstruction of the signal amplitude of the CMS electromagnetic calorimeter. Eur Phys J C46S1:26–35
2. Amr R, Hindawi SK (2013) Applying artificial neural network Hadron–Hadron collisions at LHC, artificial neural networks—architectures and applications. In: Suzuki K (ed), InTech. doi:10.5772/51273 ISBN: 978-953-51-0935-8
3. Anderson K et al (2005) Design of the front-end analog electronics for the ATLAS tile calorimeter. Nucl Instrum Methods Phys Res 551(2–3):469–476
4. Banerjee S et al (2012) CMS simulation software. J Phys Conf Ser 396
5. Behrens U et al (1994) Calibration of the forward and rear ZEUS calorimeter using cosmic ray muons. Nucl Instrum Methods Phys Res A 339(3):498–510
6. Bertuccio G, Gatti E, Sapietro M (1992) Sampling and optimum data processing of detector signals. Nucl Instrum Methods Phys Res A 322:271–279
7. Bondon P, Benidir M, Picinbono B (1992) A nonlinear approach to estimate the amplitude of a signal. IEEE Int Conf Acust Speech Signal Process 5:301–304
8. Bos A (2007) Parameter estimation for scientists and engineers, 1st ed. Wiley-Interscience, New York
9. Carrio F et al (2014) The sROD module for the ATLAS tile calorimeter phase-II upgrade demonstrator. J Instrum 9:C02019
10. Chapman J (2011) ATLAS simulation computing performance and pile-up simulation in ATLAS. LPCC detector simulation workshop, CERN
11. Cleland W, Stern E (1994) Signal processing considerations for liquid ionization calorimeters in a high rate environment. Nucl Instrum Methods Phys Res A 338:467–497
12. Collaboration CMS, Electromagnetic CMS (2011) Calorimeter status and performance with the first LHC collisions. J Phys Conf Ser 293:012042
13. Cottingham W, Greenwood D (1998) An Introduction to the Standard Model of Particle Physics. Cambridge University Press, Cambridge
14. Denby B (1999) Neural networks in high energy physics: a ten year perspective. Comput Phys Commun 119(23):219–231
15. Drake G et al (2002) The upgraded CDF front end electronics for calorimetry. IEEE Trans Nucl Sci 39(5):1281–1285
16. Evans L, Bryant P (2008) (eds) LHC Machine. J Instrum 3:S08001
17. Filho L et al (2015) Calorimeter response deconvolution for energy estimation in high-luminosity conditions. IEEE Trans Nucl Sci 39(5):3265–3273
18. Fullana E et al (2006) Digital signal reconstruction in the ATLAS hadronic tile calorimeter. IEEE Trans Nucl Sci 53(4):2139–2143
19. Gabaldon G (2009) Electronic calibration of the ATLAS LAr calorimeter and commissioning with cosmic muon signals. J Phys Conf Ser 160

20. Gleyzer SV, Prosper H (2008) An artificial neural network based algorithm for calorimetric energy measurements in CMS. In: Proceedings of XII Advanced Computing and Analysis Techniques in Physics Research, p 91
21. Hagan M, Menhaj M (1994) Training feed-forward networks with the Marquardt algorithm. IEEE Trans Neural Netw 5(6):989–993
22. Haykin S (1998) Neural networks: a comprehensive foundation. Prentice Hall, New Jersey
23. Herr W, Muratori B (2003) Concept of luminosity. CERN accelerator school: intermediate course on accelerator physics, CERN, pp 361–378
24. Hong Ma H et al. (2015) Upgraded trigger readout electronics for the ATLAS LAr calorimeters for future LHC running. J Phys Conf Ser 587:012019
25. Kay S (1993) Fundamentals of statistical signal processing, estimation theory. Prentice Hall, New Jersey
26. Labati R, Genovese A, Piuri V, Scotti F (2012) Low-cost volume estimation by two-view acquisitions: A computational intelligence approach. In: International Joint Conference on Neural Networks (IJCNN), pp. 1–8
27. Labati R, Genovese A, Piuri V, Scotti F (2012) Weight estimation from frame sequences using computational intelligence techniques. In: IEEE international conference on computational intelligence for measurement systems and applications (CIMSA), pp. 29–34
28. Livingston MS, Blewett J (1969) Particle accelerators. McGraw-Hill, New York
29. Machado V et al (2011) A new proposal to provide estimation of qos and qoe over wimax networks: an approach based on computational intelligence and discrete-event simulation. In: IEEE Latin-American conference on communications (LATINCOM), pp 1–6
30. Madsen P (1994) Neural network for combining linear and non-linear modelling of dynamic systems. IEEE World Congress Comput Intell 7:4541–4546
31. Marshall Z et al (2014) Simulation of pile-up in the ATLAS experiment. J Phys Conf Ser 513:022024
32. Meyer-Baese U (2007) Digital signal processing with field programmable gate arrays. Springer, Heidelberg
33. Nicholson PW (1974) Nuclear electronics. Wiley, New York
34. Pincibono B, Duvault P (1988) Optimal linear-quadratic systems for detection and estimation. IEEE Trans Inf Theory 34:304–311
35. Pincibono B, Chavalier P (1995) Widely linear estimation with complex data. IEEE Trans Signal Process 43:2030–2033
36. Ruggiero F (2004) LHC accelerator R&D and upgrade scenarios. Eur Phys J C Part Fields 34:433–442
37. Sarajedini A, Hecht-Nielsen R, Chau P (1999) Conditional probability density function estimation with sigmoidal neural networks. IEEE Trans Neural Netw 10(2):231–238
38. Seixas J (1999) Using neural networks to learn energy corrections in hadronic calorimeters. In: Scientific applications of neural nets, Lecture notes in physics, vol 522, pp 170–188
39. Silva PVM, Seixas JM (2001) A hybrid training method for neural energy estimation in calorimetry. AIP Conf Proc 583:86–88
40. Stoica P, Hongbin L, Jian L (2000) Amplitude estimation of sinusoidal signals: survey, new results, and an application. IEEE Trans Signal Process 48(2):338–352
41. Tang F et al (2015) Upgrade analog readout and digitizing system for atlas tilecal demonstrator. IEEE Trans Nucl Sci 62:1045–1049
42. Teodorescu L (2008) Artificial neural networks in high-energy physics. In: Inverted CERN School of Computing, CERN-2008-002, pp. 13–21
43. The ATLAS Collaboration (2010) Readiness of the ATLAS tile calorimeter for LHC collisions. EPJC 70:1193–1236
44. The ATLAS Collaboration (2010) Readiness of the ATLAS liquid argon calorimeter for LHC collisions. EPJC 70:723–753
45. The ATLAS Collaboration (2012) Observation of a new particle in the search for the standard model higgs boson with the ATLAS detector at the LHC. Phys Lett B 716:1–29

46. The ATLAS collaboration (2014) A neural network clustering algorithm for the ATLAS silicon pixel detector. J Instrum 9:P09009
47. Torres RC, Lima DEF, Filho EFS, Seixas JM (2009) Neural online filtering based on pre-processed calorimeter data. In: 2009 IEEE nuclear science symposium conference record (NSS/MIC), pp. 530–536
48. Vizireanu DN, Halunga SV (2012) Simple, fast and accurate eight points amplitude estimation method of sinusoidal signals for DSP based instrumentation. J Instrum 7
49. Wang D et al (2015) Readout electronics upgrade on ALICE/PHOS detector for Run 2 of LHC. J Instrum 10:C02025
50. Wang Z et al (2008) Application of BP neural networks in non-linearity correction of optical tweezers. Front Electr Electr Eng 3(4):475–479
51. Weng JF, Leung SH (2000) Nonlinear RLS algorithm for amplitude estimation in class A noise. IEE Proc Commun 147(2):81–86

Chapter 6
Non-supervised Learning Applied to Analysis of Topological Metrics of Optical Networks

Danilo R.B. de Araújo, Joaquim F. Martins-Filho and Carmelo J.A. Bastos-Filho

Graphs can be used to model many real-world problems, such as social networks, telecommunication networks and biological structures. To aid the analysis of complex networks, several topological metrics and generational procedures have been proposed in the last years. This work offers a systematic method to analyse different backbone optical networks, based on a non-supervised algorithm for clustering and investigates the power of a recently proposed topological metrics, named $I(\hat{\mathcal{F}})$. The metrics $I(\hat{\mathcal{F}})$ and three others are applied to identify the best canonical model to represent real backbone optical networks. According to the obtained results, the clustering procedure allows to indicate $I(\hat{\mathcal{F}})$ as the better metrics to explain the installed capacity for the analysed networks.

6.1 Introduction

Network Science is an interdisciplinary area that studies the behaviour of complex networks present in different application domains, such as telecommunication networks, biological networks, neural networks, social networks among others. The

D.R.B de Araújo (✉)
Federal Rural University of Pernambuco, Recife, Pernambuco, Brazil
e-mail: danilo.araujo@ufrpe.br

J.F. Martins-Filho
Federal University of Pernambuco, Recife, Pernambuco, Brazil
e-mail: jfmf@ufpe.br

C.J.A Bastos-Filho
University of Pernambuco, Recife, Pernambuco, Brazil
e-mail: carmelo.filho@upe.br

© Springer International Publishing Switzerland 2017
N. Nedjah et al. (eds.), *Designing with Computational Intelligence*,
Studies in Computational Intelligence 664,
DOI 10.1007/978-3-319-44735-3_6

National Research Council defines network science as the study of networks to represent physical, social and biological phenomena and the creation of models to forecast these phenomena [9]. Significant advances in network science are related to the proposition of generative procedures to create graphs with topological properties similar to the properties usually exhibited by in real-world networks. In 1940, Erdos and Renyi developed relevant studies related to random networks [12]. In 1988, Watts and Strogatz proposed the first generative procedures for networks that present the small-world effect (SW) [23]. In 1999, Barabási and Albert presented a model based on preferential attachment to generate scale-free networks (SF) [6]. Several variants of these models were already proposed in the last years [19]. In most of the cases, real-world networks do not present a topology with random properties, such as the networks generated by the Erdos–Renyi (ER) model. In general, real-world networks exhibit characteristics similar to ones observed in regular networks, small-world networks, scale-free networks or a mix of the properties from these models. Based on this, the topological properties can be used to classify networks according to a given family of graphs.

Besides, studies from European and American telecommunication backbone networks indicate that there is a high correlation between economic and demographic aspects and the optical fibre topology deployed to serve for a given traffic demand [20–22]. Moss and Townsend [20] studied the development of the Internet in the United States between 1977 and 1999 and established that there exists a correlation between the physical and logical topology of the transport networks with the presence of companies that have information as their primary asset. More recently, Tranos and Gillespie [22] studied the factor that drive the spatial distribution of the transport networks in Europe and they concluded that several variables affect the physical topology, but in general, the nodes are more connected in large metropolitan regions. Tranos [21] observed in Europe that the infrastructure of aviation networks are scale-free, whereas the backbone infrastructure is more regular. Cardenas et al. [8] presented a study in which they show that the node degree distribution of the SDH networks owned by the Telefonica-Spanish also is a power law curve. Knight et al. [17] performed a study on the emergence of canonical models in physical topologies of the Internet from several countries. This study was not conclusive about the more suitable model to represent a given transport network since some of the analysed networks present power law distribution, while others are more regular.

The studies on topological analysis of backbone networks frequently use several metrics aiming at identifying the best canonical model to explain the analysed backbone. However, there is no conclusive previous study offering a systematic approach to cluster backbone networks considering the canonical model that best fits the backbone networks. This paper proposes the use of a well-known non-supervised algorithm, called K-means [16], to group backbone networks according to the topological properties. This procedure allows one to identify the most promising features to suggest canonical models to better represent the deployed backbone networks.

The remainder of this chapter is organised as follows: Sect. 6.2 provides a short literature review on network science and clustering algorithms; Sect. 6.3 shows the proposed methodology to analyse data. Section 6.4 presents and discusses the obtained results. Finally, Sect. 6.5 gives the conclusions and suggestions for future investigations.

6.2 Theoretical Background

6.2.1 Network Science

This section aims to provide theoretical foundations related to topological metrics of networks and generative procedures to create network topologies with specific properties. A complete reference to Network Science can be found in [19].

Topological Properties of Networks

In this work, we consider a network as a graph $G = (\mathcal{N}, \mathcal{E})$, in which $\mathcal{N}$ and $\mathcal{E}$ are the set of vertices and the set of edges, respectively. The amount of nodes and links in a network are $n = |\mathcal{N}|$ and $e = |\mathcal{E}|$, respectively. In this chapter, the graphs used to represent networks are unweighted and undirected. We assumed this because the maximum number of optical channels for each optical link is fixed and it is equal in all links in the network. Besides, each connection between each pair of nodes has a pair of optical fibres, one for transmission and another one for reception.

A typical representation for G is the adjacency matrix A. The elements of the matrix A indicate a link between the nodes i and j if the element $a_{i,j} = 1$ and indicate the absence of connectivity if $a_{i,j} = 0$. If the network G contains only bidirectional connections (the case studied in this work), the A matrix is symmetric.

The node degree (d) represents the number of links that connects the node to the neighbours nodes. The node degree distribution of G defines the likelihood $Pr(d)$ of a node randomly selected in the network to have a specific node degree d. The entropy of a network $I(G)$ is calculated using the node degree distribution and provides a measure of randomness of the network.

The shortest path (SP) describes the minimum number of hops between a pair of nodes. The average path length ($\overline{c}$) is the average of the SPs considering every source–destination pair in the network. The clustering coefficient (c_i) of a node i is the rate between the number of triangles that contains the node i and the number of possible triangles if all neighbours of node i were connected. The clustering coefficient of the entire network (CC) is the average value of c_i considering all nodes of the network. The assortativity coefficient ($-1 \leq r \leq 1$) provides a measure to evaluate if nodes with similar degree are connected ($r \geq 0$). If $r \leq 0$, it is more likely to have links between nodes with different degrees. $Pr(d)$ and the other measures presented in this section are frequently used in order to specify which canonical model better represents a network of the real world.

In the graph theory, a network can also be analysed by its adjacency matrix (A) or the Laplacian matrix (L). The *node degree matrix* (D) is used together with A to build L. In this case, $L = D - A$, in which the non-diagonal entries (i, j) are either "-1" or "0", depending on whether nodes i and j are adjacent or not, respectively, and the diagonal entries (i, i) are equal to the degree of the nodes D_i. The ordered set of n eigenvalues of A or L is referred as the spectrum of the matrix. The second smallest eigenvalue of L is denoted as the *algebraic connectivity* (λ_{N-1}). A graph is disconnected if $\lambda_{N-1} = 0$. Moreover, if $\lambda_{N-i+1} = 0$ and $\lambda_{N-i} \neq 0$, then a graph has exactly i components [13]. Another important spectral metrics, called natural connectivity (NC) [24], aims to characterise the redundancy of alternative routes in a network. Both λ_{N-1} and NC are commonly used to measure the robustness of real-world networks. More recently, Araújo et al. [2] proposed a novel metrics $I(\hat{\mathcal{F}})$ based on the entropy of the Fourier coefficients of the eigenvalues of L. According to [2], $I(\hat{\mathcal{F}})$ can provide more information regarding the network structure than other metrics. $I(\hat{\mathcal{F}})$ is detailed in the next subsection.

The Topological Metrics $I(\hat{\mathcal{F}})$: Recent works demonstrated that the use of the discrete Fourier transform (DFT) over the Laplacian eigenvalues can help to identify characteristics not shown by the original set of eigenvalues. Araújo et al. [1] proposed the use of particular points of the DFT curve to classify networks according to a given canonical model. However, the metrics proposed in [1] can be used only to analyse sparse networks. The metrics $I(\hat{\mathcal{F}})$, proposed by Araújo et al. [3], is based on the entropy of the coefficients of the DFT of the Laplacian eigenvalues and can be used for sparse and dense networks. $\mathcal{L}$ provides a summary of the network topology since it contains information regarding the degrees of the nodes and the established links. Thus, a metrics obtained by the eigenvalues of $\mathcal{L}$ can correctly summarise the network topology. This fact was already exploited to propose several topological metrics [19], but most of them are not suitable for a large spread of link densities, i.e. the classification rule does not hold for sparse and dense networks.

Thus, it is possible to use the entropy of the DFT of the Laplacian eigenvalues to classify graphs according to the topology and the degree of randomness of the network. One can calculate the metrics according to the Algorithm 8. The Eq. (6.1) summarises the numerical calculation of the measure.

Algorithm 8 The algorithm for $I(\hat{\mathcal{F}})$: Procedure *CalculateIF*.

1: Let A the adjacency matrix of a graph G
2: Calculate the degree matrix D
3: Calculate the Laplacian matrix $L = D - A$
4: Calculate the real eigenvalues of L and store it in E
5: Calculate the Discrete Fourier Transform (*DFT*) over E and store values in $\mathcal{F}$
6: Normalize the $\mathcal{F}$ set in order to obtain values between 0 and 1 and store it in $\hat{\mathcal{F}}$
7: Calculate the entropy of $\hat{\mathcal{F}}$ values using Eq. (6.1)

$$I(\hat{\mathcal{F}}) = -\sum_{i=1}^{|\hat{\mathcal{F}}|}(\hat{\mathcal{F}}_i \cdot \log_2 \hat{\mathcal{F}}_i). \tag{6.1}$$

Equation (6.1) has a structure similar to the one presented by the traditional entropy used in information theory. It also looks similar to the entropy of the node degrees (well known in the network science literature). A metrics based on entropy evaluates the amount of information, the level of uncertainty and the predictability of the numerical value of some physical measure. However, the deployment of an entropy concept to the normalised set of $\mathcal{F}$ brings the uncertainty of the node degrees and the assortativity of the network. It considers simultaneously: (i) the mechanism of linking nodes by the Laplacian matrix (this allows to obtain the idea of assortativity); and (ii) the assessment of randomness by considering that entropy also acts over the node degrees, present in the main diagonal of the Laplacian matrix.

The same motivation could be directly applied to the set of eigenvalues without the necessity of an additional transform calculation (i.e. *DFT*). However, the *DFT* provides an ease of analysis of the metrics over a large range of densities and the direct use of the entropy over the normalised set of eigenvalues does not offer the same interpretation when considering different network densities. Figure 6.1 illustrates the use of three different measures of entropy *versus* link densities, for networks with 100 nodes that were generated by different canonical models. Figure 6.1 reinforces the arguments presented to justify the use of the $\mathcal{F}$. For instance, for the entropy of node degrees, presented in the Fig. 6.1a, there are different classification rules for $q < 0, 10; 0, 10 < q < 0, 96$; and $q > 0, 96$. Besides, the entropy of the eigenvalues of $\mathcal{L}$, presented in Fig. 6.1b, do not provide a unique classification rule for all the ranges of density.

The metrics $I(\hat{\mathcal{F}})$ was used in [5] and in [4] to aid the process of estimate blocking probability of optical networks. Another study has used the metrics $I(\hat{\mathcal{F}}), CC, r$ and $\bar{c}$ to analyse five different Brazilian backbones [2]. However, all those studies applied $I(\hat{\mathcal{F}})$ to analyse a small set of scenarios. This work proposes a systematic methodology to evaluate the capacity of explanation of topological metrics and applies this methodology to a set of 107 deployed optical networks of several countries.

Generational Methods to Create Networks

Random networks were the first intensely studied networks. However, random networks are not very useful to explain real network phenomena, since real-world networks always present some structure. However, random networks are often used as a benchmark to evaluate the randomness degree of real networks. Random networks show a high value for the entropy of the node degree and the clustering coefficient is small when compared to structured networks. The assortativity coefficient for random networks is approximately zero. The methods of *Erdos–Renyi (ER)* and Gilbert are the two more popular generational methods to create random networks. The ER model creates networks by including new random links until the target density is achieved. The proposal of Gilbert starts with a fully connected network and removes random links until the target density is achieved. The $Pr(d)$ of a random network

Fig. 6.1 Different entropy
measures for scale-free
networks, random networks
and small-world networks,
with 100 nodes

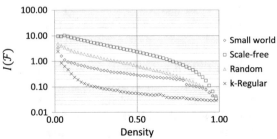

(a) Entropy of the DFT of the Laplacian eigenvalues.

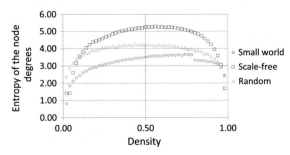

(b) Entropy of the node degrees.

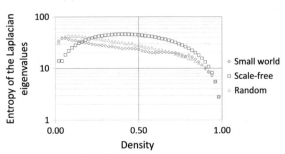

(c) Entropy of the Laplacian eigenvalues.

follows a Poisson distribution for large networks and follows a Binomial distribution
for small networks [19].

k-Regular networks are created by connecting each node i to the other nodes
$j \in \{i + 1, \ldots, i + k\}$, for $k \geq 2$. If $k = 1$, then the network is known as a ring
network. k-Regular networks are highly structured networks since all nodes have the
same degree $d = 2k$. Thus, $Pr(d)$ for k-Regular networks is a Delta function. The
entropy of the node degree is zero for k-Regular networks. A k-Regular network is
sparse if k is small and is dense for a large value for k. The density of a k-regular
is given by $\frac{2k}{n-1}$. Besides, k-regular networks are connected and present a small
value for both diameter and average path length. The assortativity coefficient for

k-Regular networks is equal to one. k-Regular networks can be used to approximate the characteristics of some real networks.

Small-world networks present a high value for the clustering coefficient and a small value for the average path length. The entropy of the node degree can be tuned by adjusting the parameters of the models. Small-world networks can be classified between regular networks and random networks. Several models were proposed to generate the small-world effect in networks and one of the most popular is the *Watts–Strogatz (WS)* method. WS method is initialized by creating a 2-Regular network. After this, one needs to establish *pm* links, in which *p* is called as the rewiring probability. The rewiring of links adds a limited amount of randomness and allows the emergence of the small-world effect related to the shortcuts. Shortcuts provide a way to decrease the diameter and the average path length of 2-regular networks. After the rewiring process, the diameter and the average path length are closer to the values obtained for random networks, but the clustering coefficient and the entropy are similar to the values obtained for regular networks. In general, the node degree distribution of small-world networks can be approximated by a Poisson function. Some previous studies established relationships between the small-world model and several real-world networks, such as social networks [19].

A scale-free network presents a node degree distribution that obeys a power law $h(d) \sim d^{-q}$, in which d is the degree and q is an exponent often in the range [2, 3]. A network that follows this model presents few nodes with many links (known as hubs), but most of the nodes has just one or some few connections. Scale-free networks show a high value for the entropy, but below the values for equivalent random networks. The diameter and average path length of scale-free networks are small due to the presence of hubs. The *Barabási–Albert (BA)* method is the most popular procedure to generate scale-free networks. The BA model initiates with a fully connected graph with three nodes. Then, the other nodes are connected to the network by using the preferential attachment mechanism. This means that each arriving node is connected to the others according to a probability $P(d_i) = \frac{d_i}{\sum_{j=1}^{n} d_j}$, i.e. the arriving nodes have a higher tendency to connect to nodes with higher degree. Because of this, the assortativity coefficient for scale-free networks presents a negative value. Previous studies demonstrated that several real-world networks, such as Internet, railroads and biological systems, follows this model [19]. Dougherty et al. [11] proposed a generalised form for the preferential attachment model in which the attractivity of a node is given by

$$U(s, t) = M(s, t)k^{\tau}(s, t) + S(s, t). \tag{6.2}$$

wherein U, M and S are functions that depend on the node s and time t. M is the fitness value of node s at time t, and S is an additional attractiveness of node s at time t. If we consider the generalised form, the probability for an arriving node to be attached to an existing node i at time t is given by

$$P(i, t) = \frac{U(i, t)}{\sum_{j=1}^{n} U(j, t)}. \tag{6.3}$$

Table 6.1 Comparison of some topological properties for networks with 50 nodes, 100 edges and $rp = 5\%$

Property	ER	BA	WS	2-regular
Assortativity	−0.06	−0.24	0.07	1.00
CC	0.07	0.14	0.44	0.50
Distribution	Poisson	Power	Poisson	Delta(4)
Maximum degree	9	19	5	4
Average degree	4.0	3.96	4.0	4
$\bar{c}$	2.83	2.68	4.54	6.63
$I(\hat{\mathcal{F}})$	2.5	3.54	1.0	0.82

Using this generalised preferential attachment model, different variants of the scale-free networks can be created. For example, if $\tau = 1$, $M(s, t) = 1$ and $S(s, t) = 0$, then the generalised model becomes the BA model.

Table 6.1 shows numerical examples of topological properties of 50-node networks built by each one of the models we have presented in this section.

6.2.2 Clustering Algorithms

A clustering algorithm takes a set of data points and group the points into clusters (subsets). Several algorithms have already been proposed to group data. In this section, we provide the main characteristics of several approaches. For a more broad study about these algorithms, we recommend several references.

- K-means: it is one of the most common iterative algorithms [16], broadly used because of its simplicity of implementation, its convergence speed and the sound quality of the clusters (for a limited family of problems). In the K-means algorithm, each vector is classified as belonging to a cluster, and the centroids are updated based on the classified samples, according to some distance metrics (such as the Euclidean distance).
- Fuzzy K-means: in the K-means algorithm, each vector is classified as belonging to a unique cluster (hard cluster). In a variation of this approach, known as fuzzy C-means [16], all vectors have a degree of membership to belong to each cluster, and the respective centroids are calculated based on these membership degrees.
- SOM (Self-Organising Map): by applying an approach known as self-organizing map, the clusters can be defined by the points of a grid adjusted to the data [18]. Usually, the algorithm uses a two-dimensional grid in the higher-dimensional space, but it is usual to use a one-dimensional grid for clustering purposes.
- Hierarchical clustering (HC): creates a hierarchical tree of similarities between the vectors, called dendrogram. The most common implementation of HC is the agglomerative hierarchical clustering, which starts with a family of clusters with

one vector each, and merges the clusters iteratively based on some distance measure until there is only one cluster left, containing all the vectors [16].

In this work, we adopt the K-means algorithm due to its simplicity and because the first results obtained for our problem by this algorithm were satisfactory.

Assessment of Clustering Approaches

Despite the popularity of clustering, until very recently little attention has been paid to measure the output of a clustering algorithm.

Quality indicators proposed for clustering algorithms can be classified into three types. The first type is based on calculating properties of the resulting clusters, such as compactness, separation and roundness. This approach is called internal validation, because it does not require additional information about the data [14, 15]. The second approach is based on comparisons of the partitions generated by the same algorithm with different parameters, or different subsets of the data. This procedure is called relative validation, and also does not include additional information [7, 15]. In the third possibility, called external validation, the approach is also based on the comparison of the partitions. The partitions consist of the one generated by the clustering algorithm and a given partition of the data (or a subset of the data) [7, 11]. External validation corresponds to a kind of error measurement, either directly or indirectly. Therefore, we should expect external methods to be better correlated to the true error. However, this is not always the case since it depends on the external validation procedure. It also depends on the random labelled point process being applied and the particular clustering algorithm being tested.

In this work, we used a simple external validation indicator when labelled data are available, i.e. when we use synthetic networks generated by the canonical model. According to Brun et al. [7], based on their extensive simulations among varied models, the silhouette indicator almost always outperforms the other internal indices. Thus, we used the silhouette indicator as an internal measure to analyse our experiments. The higher the silhouette, the more compact and separated are the clusters. The silhouette of a vector x of a cluster C_k is defined by the Eq. (6.4):

$$S(x) = \frac{b(x) - a(x)}{max[b(x), a(x)]},$$

(6.4)

wherein $a(x)$ is the average distance between x and all the other patterns in C_k and $b(x)$ is the minimum of the average distances between x and the vectors of the other clusters. $S(X)$ ranges from -1 to $+1$. If the value is close to -1, then this means that the vector is closer, on average, to another cluster than the one to which it belongs. If the value is close to $+1$, then it means that its average distance to its cluster is significantly smaller than to any other cluster. The global *silhouette index* is given by the Eq. (6.5):

$$S = \frac{1}{K} \sum_{k=1}^{K} \left[\frac{1}{n_k} \sum_{x \in C_k} S(x) \right].$$

(6.5)

6.3   Our Methodology

The studied problem in this chapter can be described as follows: given a set of topological properties of networks with different characteristics, the goal is to cluster the topologies according to the canonical model that best fits each network. In this work two canonical models are considered, i.e. the Barabási–Albert (BA) model and the Watts–Strogatz (WS) model. Graphs generated by the BA model represent typical *scale-free* networks and the graphs produced by the WS model present the *small-world* effect as the main characteristic [19]. We generated several networks using both models and next, those networks are clustered by using the K-means algorithm [16]. The K-means allows the clustering of patterns in k groups according to the previously chosen properties. Besides, several topological metrics are used to evaluate the explanation capacity of each one of those metrics regarding the best canonical model to represent the networks. Our implementation of the K-means considers that the centroids are randomly initialized and the algorithm stops when the centroids do not change in two subsequent iterations.

We considered different average degrees $(\overline{d})$ for the networks generated by canonical models. When we take into account the clusters to group sparse and dense networks and to group the networks generated by the BA and WS models, we think it is necessary at least $k = 4$ clusters. The topological metrics used in the non-supervised learning process are: average node degree $\overline{d}$, entropy of the node degrees $I(G)$, assortativity coefficient r, degree of the largest connected node $d(h)$ and the entropy of the DFT of the Laplacian eigenvalues $I(\hat{\mathcal{F}})$. To identify the capacity of a set of metrics to correctly cluster the networks according to the canonical model, we adopt some definitions and procedures for the external analysis. For the internal analysis, we used the silhouette indicator shown in the Sect. 6.2.2. We define *preferential canonical model* as the model related to the generative procedure that creates the network that is closer to the centroid of each cluster found by the K-means. We define the *success rate of a cluster* as the percentage of networks of a cluster that are generated by the preferential canonical model of the considered cluster. We define the *capacity of explanation* of a set of metrics as the average of the success rate for the whole set of K clusters. One of the goals of this study is to identify the best set of topological metrics regarding capacity of explanation to discover which is the canonical model that best fits a given real-world network.

Our study is divided into two main parts. In the first part, we generated several physical topologies using canonical models. In this first step is considered the generation of different topologies for a 32-node network, which is the average number of nodes found in our dataset of deployed networks. We consider 500 different networks for each one of the canonical models and we use the average degree between 2.0 and 5.0. Thus, the clustering procedure considers a set of 1000 networks. The WS is set to use a rewiring probability $rp = 0.10$ and the k parameter of the regular network used at the beginning of the procedure is chosen according to the specified average degree. The BA model also considers an additional number of links Δm proportional to the average degree determined for each network. The second part of the study consists in the analysis of the clustering procedure for a dataset with 107 optical networks deployed in different countries. The primary goal of the first step is to understand the capacity of explanation of the metrics when the canonical model is well-known (labelled data). The main objective of the second step is to understand the general topological profile of the networks contained in a large dataset of real optical networks and to check if the clustering algorithm and the metrics defined in the first step can be applied to a real scenario (unlabeled data).

6.3.1   Data Collection and Analysis Tools

The physical topologies of the deployed networks were obtained from maps and tables available both on printed data and in the Internet. We have selected a few Brazilian national backbones available in [10] and several deployed optical networks around the world available in [17]. We used optical networks from the original datasets that are related to connected graphs and that present more than 12 nodes. The next step after the collection of raw data was to build manually for each network a structured file in the GML format (graph modelling language). The GML format contains elements to represent the nodes, links and additional information regarding the optical network, such as the geographical coordinates of the nodes and the capacity of links. By using the GML file as an input, it is possible to calculate several useful metrics to analyse the physical topology.

We built a platform to examine complex networks written in the Java programming language to allow the calculation of the topological metrics. This tool also enables the generation of several networks using canonical models. Besides, it is also possible to use the GML file to create geolocated graphs, i.e. it is possible to create maps with the graph plotted in the correct geographical coverage. To help in this process, we used the JXMapViewer API. Our dataset is now available on the Internet.[1]

6.4   Results

Table 6.2 presents the summary of the clustering process obtained by the K-means for 1000 network topologies that were generated by the BA and WS canonical models and for a 32 node network. All the rows of Table 6.2 uses the average degree measure $(\overline{d})$ to allow a separation between sparse and dense networks. The remaining metrics are used to enable the clustering of network topologies according to the canonical model. From the results, it is possible to notice that the clustering by $\overline{d}$ and $I(\hat{\mathcal{F}})$ provide the highest capacity of explanation when it is compared with the other choices of two

---

[1] www.researchgate.net/publication/283727923_dataset_optical_networks_20151113?ev=prf_pub.

Table 6.2 Results of the clustering procedure obtained by the K-means for different sets of topological properties

Number of Metrics	Metrics	Capacity of expl.		Silhouette index	
		Mean	Std. dev.	Mean	Std. dev.
2	$\overline{d}, r$	0.9311	0.0405	0.5219	0.0336
	$\overline{d}, d(h)$	0.8976	0.1617	0.6045	0.0473
	$\overline{d}, I(G)$	0.9432	0.0942	0.5653	0.0179
	$\overline{d}, I(\hat{F})$	0.9461	0.0514	0.5683	0.0595
3	$\overline{d}, I(\hat{F}), r$	0.9907	0.0288	0.5267	0.0423
	$\overline{d}, I(\hat{F}), d(h)$	0.9902	0.0040	0.5615	0.0941
	$\overline{d}, I(\hat{F}), I(G)$	0.9849	0.0270	0.5638	0.0632
4	$\overline{d}, I(\hat{F}), I(G), r$	0.9954	0.0088	0.4872	0.0758
	$\overline{d}, I(\hat{F}), I(G), d(h)$	0.9925	0.0355	0.5330	0.0830
5	$\overline{d}, I(\hat{F}), I(G), d(h), r$	0.9969	0.0069	0.4738	0.0957

variables. The best clustering in terms of capacity of explanation occurs when all the metrics are used, i.e. for the set $\{\overline{d}, I(\hat{F}), I(G), d(h)\}$. This best set of five metrics provides an average capacity of explanation equal to 0.9969. The improvement on the explanation capacity when several rows of Table 6.2 are compared indicates that the more important metrics to represent the canonical models are, in ascending order: $I(\hat{F})$, $I(G)$, r and $d(h)$. According to results shown in Table 6.2, it is possible to conclude that the best set of variables to separate networks with different densities and according to the BA and WS canonical models are: $\overline{d}$ and $I(\hat{F})$, when two metrics are used; $\overline{d}$, $I(\hat{F})$ and r, when three metrics are used; and $\overline{d}$, $I(\hat{F})$, $I(G)$ and r, when four metrics are used. Besides, the improvement regarding capacity of explanation when the clustering uses five metrics instead two metrics is only 5 %. Thus, the set $\overline{d}$ and $I(\hat{F})$ can be viewed as a good choice to indicate the canonical model of a real network when a simple mechanism of comparison is needed. The standard deviation of the capacity of explanation is less than 4 % when more than three metrics is used and this reinforces the confidence of the obtained results.

The results for the silhouette indicator suggests that the clustering procedure was successful regarding compactness and separation since it has obtained $S > 0, 47$ for all the rows of the Table 6.2. Individual values of the silhouette are between -1 and 1 and a positive value indicates that the patterns are located in a suitable cluster. However, the correlation between the silhouette indicator and the external evaluation, regarding the capacity of explanation, falls in some cases. For instance, S is very high for the set $\{\overline{d}, d(h)\}$, because it is easy to separate networks according to the presence of hubs. However, the capacity of explanation of this set is small because it is not possible to use just these variables to identify the canonical models. Thus, we can use S only as an indicator that the clustering procedure is working, but we can not ensure that each set of metrics is the best choice to identify the canonical models.

Table 6.3 Examples of clusters found by different sets of topological properties

Metrics	Cluster 1			Cluster 2			Cluster 3			Cluster 4		
	BA	WS	Centroid	BA	WS	Centroid	BA	WS	Centroid	BA	WS	Centroid
$\bar{d}, r$	3	171	2.00; 0.20	248	14	2.56; −0.30	288	59	4.13; −0.23	11	296	3.88; 0.13
$\bar{d}, d(h)$	35	185	2.00; 4.00	252	0	2.75; 12.00	221	0	4.31; 16.00	32	355	3.88; 6.00
$\bar{d}, I(G)$	222	0	2.56; 2.20	25	186	2.00; 0.88	293	6	4.13; 2.77	10	348	3.88; 1.75
$\bar{d}, I(\hat{F})$	7	185	2.00; 1.76	259	0	2.56; 2.80	254	0	4.13; 2.30	20	355	3.88; 1.28
$\bar{d}, I(\hat{F}), r$	2	182	2.00; 0.18; 1.78	268	3	2.75; −0.33; 2.76	263	1	4.13; −0.29; 2.28	7	354	3.94; 0.08; 1.25
$\bar{d}, I(\hat{F}), d(h)$	3	185	2.00; 3.00; 1.76	256	0	2.56; 11.00; 2.66	276	0	4.13; 16.00; 2.27	15	355	3.88; 6.00; 1.27
$\bar{d}, I(\hat{F}), I(G)$	0	185	2.00; 0.88; 1.76	237	0	2.56; 2.16; 2.80	299	4	4.13; 2.77; 2.33	14	351; 51	3.94; 1.77; 1.24
$\bar{d}, I(\hat{F}), I(G), r$	0	184	2.00; 0.84; 0.14; 1.76	233	11	2.44; 2.04; −0.32; 2.87	305	1	4.00; 2.79; −0.30; 2.29	2	354	4.00; 1.79; 0.10; 1.28
$\bar{d}, I(\hat{F}), I(G), d(h)$	0	185	2.00; 0.88; 3.00; 1.76	240	10	2.44; 2.17; 11.00; 2.84	299	0	4.00; 2.81; 15.00; 2.26	1	355	3.94; 1.77; 6.00; 1.24
$\bar{d}, I(\hat{F}), I(G), d(h), r$	0	185	2.00; 0.88; 0.13; 3.00; 1.77	240	10	2.44; 2.17; −0.29; 11.00; 2.84	299	0	4.13; 2.83; −0.25; 15.00; 2.46	1	355	4.00; 1.79; 0.10; 6.00; 1.28

Table 6.3 presents examples of the obtained clusters for each set of variables analysed in the Table 6.2. When we consider the average degree of the networks between 2 and 5, the average point of the groups related to sparse and dense networks has a relation to centroids with average degrees of 2.75 and 4.25, respectively. Thus, we can notice that most of the rows in Table 6.3 presents four well-defined groups related to BA sparse networks, BA dense networks, WS sparse networks and WS dense networks. For instance, the first row in Table 6.3 separates WS sparse networks in the Cluster 1, BA sparse networks in the Cluster 2, BA dense networks in the Cluster 3 and WS dense networks in the Cluster 4.

Figure 6.2 presents an analysis of the evolution of the indicators along the iterations for the K-means algorithm with the best set of metrics. Each point of Fig. 6.2 is an average of 100 different executions of the K-means algorithm in the correspondent stage of the clustering procedure. It is possible to notice that the K-means algorithm converges very fast four our problem. The capacity of explanation is close to 0.92 after the first iteration and quickly grows toward 0.99 after few iterations. Besides, the silhouette index starts close to 0.34 and also rises quickly to the asymptotic value, that is approximately equal to 0.50.

Table 6.4 presents a summary of the clustering procedure for a dataset of 107 optical networks [17], by using $k = 8$ and the variables n, q and $I(\hat{\mathcal{F}})$. Our hypothesis is that good metrics used to analyse the synthetic networks in the first step of our study are also suitable to analyse real-world networks. In this second step, we also used n since the networks of the dataset present different number of nodes. The clustering procedure for real optical networks is more challenging when it is compared with the clustering of the synthetic networks. This occurs since the deployed networks are not purely scale-free or exclusively small world. The average value of the silhouette index was around 0.30 for the clustering of these 107 networks. The profile of the networks was built by comparison of the metrics with the average value for the dataset, i.e. it is considered that a given network has few nodes when the observed amount is lower than the average value of the entire dataset (32 nodes). When we consider that $I(\hat{\mathcal{F}})$ varies according to the number of nodes and the density of the network to decide when the value is high or low, the numerical value is compared with the value obtained for an ER network with the same number of nodes and the same density, as suggested in [3].

Fig. 6.2 Average value of the metrics based on 100 different executions for each iteration of the K-means algorithm

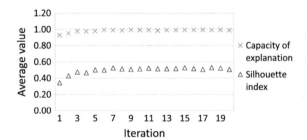

Table 6.4 Results of the algorithm k-means using n, $\bar{d}$ and $I(\hat{\mathcal{F}})$, for a database of 107 optical fibre networks [17]

Cluster	Network profile	Centroid			
		Network	n	$\bar{d}$	$I(\hat{\mathcal{F}})$
1	Few nodes, sparse, low entropy	BELNET	19	2.10	1.31
2	Few nodes, dense, low entropy	HIBERNIA	20	2.70	1.36
3	Few nodes, dense, low entropy	ATT	25	4.48	1.65
4	Few nodes, dense, large entropy	RNP	28	2.21	1.77
5	Many nodes, dense, large entropy	BICS	33	2.91	2.18
6	Many nodes, sparse, large entropy	IRIS	51	2.51	2.80
7	Many nodes, sparse, large entropy	FORTHNET	60	1.97	5.80
8	Many nodes, dense, large entropy	EMBRATEL	76	2.63	3.46

From the results presented in Table 6.4, we can notice that there are optical networks that present topological features of BA and WS networks. However, the amount of networks positioned in each group of the Table 6.4 is not uniform. For instance, the Cluster 1 has 27 networks and the Cluster 7 has only two optical networks. The clusters that have high value for $I(\hat{\mathcal{F}})$ contain 29 networks. Thus, according to the obtained results, it is possible to notice that 27 % of the networks from the analysed dataset are mainly scale-free networks that present few hubs. On the other hand, 73 % of the networks from the dataset are more regulars and do not present hubs. Figure 6.3 presents a geographical map of the optical networks related to the centroids of Table 6.4. It is possible to identify the presence of hubs in the maps presented in Fig. 6.3c–e. In the other topologies, the node degree distribution is more uniform. The American network ATT, as in Fig. 6.3c has a hub with degree equal to 9 located in Dallas and other two important hubs in Chicago and San Francisco, both with degree equal to 8. The Brazilian network RNP, as in Fig. 6.3d has a hub with degree equal to 5 in Brasilia, that has an amount of links above the average of the other cases (the average is two links per node). The FORTHNET Greek network, as in Fig. 6.3e has several hubs, but the most important is located in Athens, with 17 links. An interesting analysis is a comparison between the RNP Brazilian network, as in Fig. 6.3d and the EMBRATEL one, as in Fig. 6.3f. Although both networks have national geographical coverage, the EMBRATEL network has more nodes and a larger amount of links. However, although the $I(\hat{\mathcal{F}})$ metrics presents a higher value for the EMBRATEL than for the RNP, it is possible to notice that this is due to the additional number of links in the EMBRATEL network. Thus, it is necessary to use an ER network as benchmarking when networks with a different number of nodes are compared. The ER network related to the RNP and EMBRATEL has $I(\hat{\mathcal{F}})$ equal to 1.70 and 4.02, respectively. As the RNP presents $I(\hat{\mathcal{F}})$ above the reference value, and the EMBRATEL presents $I(\hat{\mathcal{F}})$ below the reference value; we can conclude that these networks are best represented by the BA and WS models, respectively. In the Fig. 6.3f it is possible to observe a more regular structure in the connections of the

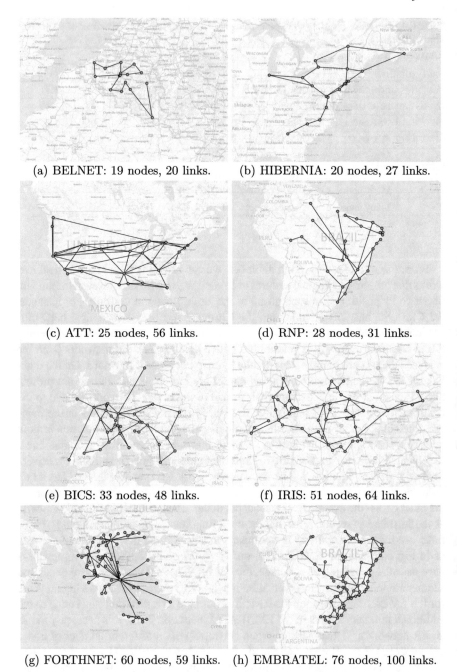

(a) BELNET: 19 nodes, 20 links.　　　(b) HIBERNIA: 20 nodes, 27 links.

(c) ATT: 25 nodes, 56 links.　　　(d) RNP: 28 nodes, 31 links.

(e) BICS: 33 nodes, 48 links.　　　(f) IRIS: 51 nodes, 64 links.

(g) FORTHNET: 60 nodes, 59 links.　　　(h) EMBRATEL: 76 nodes, 100 links.

Fig. 6.3 Optical networks related to the centroids obtained by the k-means over a dataset of 107 optical networks using the metrics n, $\overline{d}$ and $I(\hat{\mathcal{F}})$

EMBRATEL network because there are interconnected rings in the Northeast, South and Southwest regions.

We want to emphasise that the highlighted networks of the Fig. 6.3 are topologically distinct since it presents differences concerning the number of nodes, link densities and the shape of the network. Thus, the set of networks related to the centroids can be used in studies in which it is necessary to evaluate the impact of different algorithms on different network topologies, such as for the proposal of a new routing algorithm, for example.

6.5 Conclusions

In this chapter, we analysed the capacity of explanation of several topological metrics regarding the canonical model that best fits a given real network. The proposed methodology uses the non-supervised learning algorithm K-means. According to the obtained results for several installed backbone networks, the clustering procedure of the K-means provide an easy way to analyse the topological properties of the networks. Besides, the $I(\hat{\mathcal{F}})$ metrics is the best one in terms of individual capacity of explanation when it is compared with the entire set of metrics.

Future works aim to investigate other features of the backbone networks, such as the geographical coverage and the capacity in terms of traffic load balance. Besides, other canonical models that presents different topological characteristics can be applied to evaluate the proposed methodology in different scenarios. At last, other clustering algorithms can be evaluated, such as the fuzzy K-means, SOM approaches and algorithms for hierarchical clustering.

Acknowledgments The authors acknowledge the financial support from CNPq, CAPES, UFPE, UFRPE and UPE.

References

1. Araújo DRB, Bastos-Filho CJA, Martins-Filho JF (2013) Towards using DFT to characterize complex networks. In: XXXI Simpósio Brasileiro de Telecomunicacoes (SBrT2013), pp 1–5
2. Araújo DRB, Bastos-Filho CJA, Martins-Filho JF (2014) Métricas de redes complexas para análise de redes Ópticas. Revista de Tecnologia da Informacao e Comunicacao 4(2):1–8
3. Araújo DRB, Bastos-Filho CJA, Martins-Filho JF (2014) Using the entropy of DFT of the Laplacian eigenvalues to assess networks. In: Complex networks V. Studies in computational intelligence, vol 549. Springer, Heidelberg, pp 209–216
4. Araújo DRB, Bastos-Filho CJA, Martins-Filho JF (2015) Artificial neural networks to estimate blocking probability of transparent optical networks: a robustness study for different networks. In: Proceedings of the 17th international conference on transparent optical networks ICTON 2015
5. Araújo DRB, Bastos-Filho CJA, Martins-Filho JF (2015) Methodology to obtain a fast and accurate estimator for blocking probability of optical networks. J Opt Commun Netw 7(5):380–391

6. Barabási AL, Albert R (1999) Emergence of scaling in random networks. Science 286:509–512
7. Brun M, Sima C, Hua J, Lowey J, Carroll B, Suh E, Dougherty ER (2007) Model-based evaluation of clustering validation measures. Pattern Recognit 40(2007):807–824
8. Cardenas JP, Benito RM, Mouronte ML, Feliu V (2009) The effect of the complex topology on the robustness of Spanish SDH network. In: Fifth international conference on networking and services, 2009. ICNS '09, pp 86–90. doi:10.1109/ICNS.2009.28
9. Committee on Network Science for Future Army Applications, N.R.C (2005) Network science. The National Academies Press, Washington, D.C. http://www.nap.edu/openbook.php?record_id=11516
10. Cordeiro L (2013) Atlas Brasileiro de Telecomunicações. 13. Converge Comunicações
11. Dougherty E, Barrera J, Brun M, Kim S, Cesar R, Chen Y, Bittner M, Trent J (2002) Inference from clustering with application to gene-expression microarray. J Comput Biol 9(1):105–126
12. Erdos P, Rényi A (1960) On the evolution of random graphs. Publication of the Mathematical Institute of the Hungarian Academy of Sciences, pp 17–61
13. Fiedler M (1973) Algebraic connectivity of graphs. Czechoslov Math J 23:298–305
14. Guenter S, Bunke H (2001) Validation indices for graph clustering. In: Proceedings of the 3rd IAPR-TC15 workshop on graph-based representations in pattern recognition, pp 229–238
15. Halkidi M, Batistakis Y, Vazirgiannis M (2001) On clustering validation techniques. Intell Inf Syst J 17(2–3):107–145
16. Jain A, Murty M, Flynn P (1999) Data clustering: a review. ACM Comput Surv 31(3):264–323
17. Knight S, Nguyen HX, Falkner N, Bowden R, Roughan M (2011) The Internet topology zoo. IEEE J Sel Areas Commun 29(9):1765–1775. doi:10.1109/JSAC.2011.111002
18. Kohonen T (1997) Self-organizing maps, 2nd edn. Springer, New York
19. Lewis TG (2009) Network science - theory and applications. Wiley, New York
20. Moss ML, Townsend AM (2000) The Internet backbone and the American metropolis. Inf Soc J 16(1):35–47
21. Tranos E (2011) The topology and the emerging urban geographies of the Internet backbone and aviation networks in Europe: a comparative study. Environ Plan 43(2):378–392
22. Tranos E, Gillespie A (2009) The spatial distribution of Internet backbone networks in Europe: a metropolitan knowledge economy perspective. Eur Urban Reg Stud 16(4):423–437
23. Watts DJ, Strogatz SH (1998) Collective dynamics of small-world networks. Nature 393:440–442
24. Wu J, Barahona M, Tan YJ, Deng HZ (2011) Spectral measure of structural robustness in complex networks. IEEE Trans Syst, Man Cybern, Part A: Syst Hum, 41(6):1244–1252. doi:10.1109/TSMCA.2011.2116117

Chapter 7
Mole Features Extraction for a Melanoma Recognition System

Henrique C. Siqueira and Bruno J.T. Fernandes

The cancer is a painful disease that kill too many people. Skin cancer is among the most frequent types of tumors in the world, and melanoma is the most worrying type of skin cancer due to its high metastasis chances. Its global occurrence index is close to 133.000 people per year. Irresponsible exposure to the sun causes 40% out of the total. Melanoma is fatal when not diagnosed it its initial stages. The most common diagnosis method is done visually based on five features: asymmetry, border, color, diameter and elevation, also kwon as ABCDE method. We propose three algorithms to extract features of skin moles based on dermatological studies, using digital image processing techniques existing in the lecture. The first feature measures the asymmetry level of the mole; the second one calculates irregularity of the edges, and the third one computes the color variance of the mole. We also evaluate these features as input to classifiers creating a melanoma recognition model that indicates whether is melanoma or normal mole. The analysis of results are shown through ROC curve and 10-fold cross-validation from two dermatological datasets: Atlas of Clinical Dermatology and DermNet NZ.

H.C. Siqueira (✉) · B.J.T. Fernandes
Escola Politécnica, University of Pernambuco, Recife, Brazil
e-mail: hcs@ecomp.poli.br

B.J.T. Fernandes
e-mail: bjtf@ecomp.poli.br

© Springer International Publishing Switzerland 2017
N. Nedjah et al. (eds.), *Designing with Computational Intelligence*,
Studies in Computational Intelligence 664,
DOI 10.1007/978-3-319-44735-3_7

7.1 Introduction

Nowadays, skin cancer is among the most common cancers in the world, especially in tropical countries because of the high incidence of UV rays. In Brazil is the most common tumor corresponding to 25 % of all malignant tumors already registered according to research conducted by INCA (National Cancer Institute of José Alencar Gomes da Silva) [11]. Its incidence has rapidly increased approximately 3–7 % rates for people with light skin.

There are three main types of skin cancer: basal cell carcinoma (BCC), squamous cell carcinoma (SCC) and melanoma. The last one represents 4 % of disease diagnostics and its major incidence is in adults with light skins. However, it is the worst case due to its high metastasis chances, the dissemination of the cancer lesion for the other organs.

The successful treatment of this cancer increases considerably if the tumor is identified in the early stages. As evidence, there was a great improvement in survival of melanoma patients that had early diagnosis in the last years [11], and about 90 % of the cases have been completely cured when the tumors was found less than one millimeter of diameter [15].

Scientific works have been developed with the objective to create support systems for melanoma diagnosis since 1980s [3]. Nevertheless, there is not a definition about the most precise model to diagnose this pathology. One reason is the difficulty to compare the models, because are applied different statistical methods to validate, and from different databases, some of these data sets created by their authors, like in Manousaki et al. [12].

In this work, we propose three new algorithms capable to extract mole features from the human skin based on real features, defined by the medical community, using digital image processing techniques. We also propose a recognition model able to distinguish whether a mole image is a melanoma or not based on such features, using one of the following classifiers for decision-making: artificial neural network (ANN), logistic regression (LR) and support vector machine (SVM).

The chapter is organized as follows. Firstly, we general explain the melanoma disease. Thus, the features extraction algorithms are introduced, as well as the model to classify melanoma using our features as input. Finally, the experimental study is presented followed by some concluding remarks.

7.2 Melanoma

The melanoma is a type of skin cancer that affects the melanocytes cells, located in the bottom of the skin's epidermis as shown in Fig. 7.1. The melanocytes produce the melanin, the pigment responsible for the skin color.

The melanoma may begin like a mole that grows over time, may appear in almost any color (including red, blue, brown, black, gray, and tan), usually has irregular

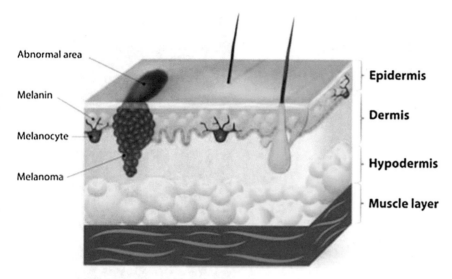

Fig. 7.1 Skin's layers with a melanocyte cell and a melanoma. Adapted from The Skin and Cancer Foundation Inc. 2016, Retrieved from https://www.skincancer.asn.au/page/2149/learn-about-melanoma

edges, may be flat or raised on skin, may be painless or form wound. His appearance is independent of anywhere of the body, but it is more common in areas exposed to the sun, such as shoulders, head, arms and legs [5].

The most common procedure for the melanoma identification is made by dermatoscopy, an examination that usually performed through the dermatoscope, handheld microscope that magnifies the skin ten times. Thus, the analysis identifies five main features, also kwown as ABCDE method (**A**symmetry, **B**order, **C**olor, **D**iameter and **E**levation) [5], as illustrated in Fig. 7.2.

- Asymmetry: indicates the level of similarity between the two halves of the mole. The Fig. 7.2a shows an asymmetrical melanoma on left and a symmetrical mole on right;
- Border: melanoma presents irregularity on the edges. In Fig. 7.2b, the top pictures show the border irregularity present in the melanoma, in opposite, the bottom pictures shows the mole with smooth transition on the edges;
- Color: melanoma has more than one color in the same mole. The Fig. 7.2c presents the histogram of the moles. In the top, melanoma case, the histogram shows the wide range of intensities, while in the bottom, has a small range of intensities.
- Diameter: melanoma is usually larger than 6 mm. Example of the melanoma case with approximately 2 cm of diameter in Fig. 7.2d.
- Elevation: It is more common to find melanoma which create a raised surface on the skin Fig. 7.2e.

Fig. 7.2 Illustration of mole
features: asymmetry (**a**),
border (**b**), color variance
(**c**), diameter (**d**) and
elevation (**e**)

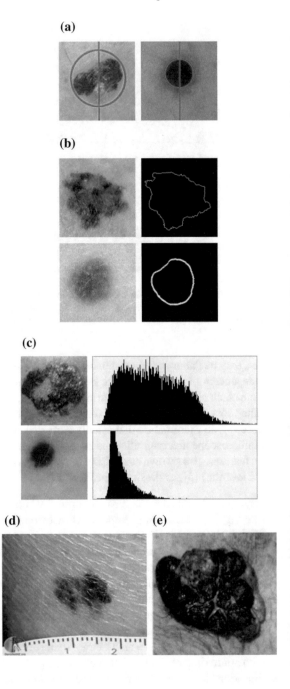

7.3 Melanoma Recognition Model

Figure 7.3 presents the complete workflow of the melanoma recognition model. The input of the model is a color image with any dimension, as illustrated in Fig. 7.4a. Next, it is necessary to identify the mole in the picture. So, we assume that image contains two classes: the first one is the mole defined as an object, and the other is the skin defined as background. The choice of the segmentation algorithm is an important decision because it contributes to the effectiveness of next steps. According to Bhuiyan et al. [2], which compares segmentation methods applied to binary images of a mole in the skin, the segmentation method by Otsu [14] achieves the best results. This method segments the grayscale image previous converted, as shown in Fig. 7.4b, in two classes as in Fig. 7.4c, based on the calculation of the optimal threshold in the image histogram that minimizes intra-class variance and maximize inter-class variance. Furthermore, does not require to set any parameter for different skin colors and moles.

Fig. 7.3 Workflow of proposed model for melanoma classification

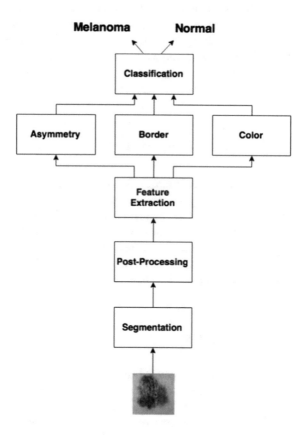

(a)   **(b)**   **(c)**   **(d)**

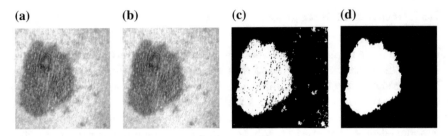

Fig. 7.4 Original image in (**a**). Converted grayscale image (**b**). The segmented image in (**c**). Image of segmented mole without fails in (**d**)

After segmentation, we make post processing to correct any fails in the binary image in Fig. 7.4d that has resulted of Otsu. Failures are skin areas that have been identified as a mole or holes present in the mole. A hole is a background region surrounded by pixels that represent the object. We used the algorithm of Suzuki et al. [16] to find the contour of all the object regions, and adopted as a mole the region with the highest number of connected components. With possession of contour points we perform the method *fillpoly()* of OpenCV [13] to fill the region, resulting in Fig. 7.4d, a segmented mole image without fails, after used as mask. Finally, the features are extracted and used as input to the supervised classifier. The output of the classifier indicates the presence of melanoma or not.

7.3.1 Feature Extraction

Three methods to extract features of the mole have been created, where we believe that they can be a good representation for the normal and the abnormal mole. The algorithms measure three melanoma characteristics previously covered: the asymmetry level, the border shape, and the color shades in the mole. The other features, diameter and elevation, are disregarded because we have not possible to get the real dimensions of the mole in the two-dimensional picture since the distance at which the image was captured is unknown.

7.3.2 Asymmetry

The asymmetry is calculated with the alignment mole by rotating the segmented image by an angle r between the mole orientation axis o and the x axis of the picture. The Fig. 7.5 illustrates the mole, the axes and the relation between angles. To obtain the r value, it is necessary to calculate the coordinates of the centroid point $p_c(l_c, c_c)$ of the mole, given by,

Fig. 7.5 Relation between
angles in the mole

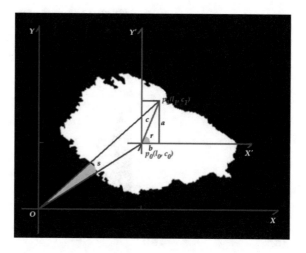

$$p_c(l_c, c_c) = \left(\frac{m_{10}}{m_{00}}, \frac{m_{01}}{m_{00}} \right) \qquad (7.1)$$

where m_{xy} are the spatial moments of order xy.

According to Horn et al. [10], to trace the orientation axis through the centroid
we have to get two points: $p_0(l_0, c_0)$ and $p_1(l_1, c_1)$ based on angle s, obtained by:

$$s = 0.5 * \arctan \left(\frac{2 * mu_{11}}{(mu_{20} - mu_{02})} \right) \qquad (7.2)$$

$$p_0(l_0, c_0) = (l_c - (100 * \cos(s))), c_c - (100 * \sin(s))) \qquad (7.3)$$

$$p_1(l_1, c_1) = (l_c + (100 * \cos(s))), c_c + (100 * \sin(s))) \qquad (7.4)$$

where mu_{xy} are the central moments of order xy. Finally, r is obtained through the
law of tangents, trigonometric formula given by,

$$r = \arctan \left(\frac{a}{b} \right) \qquad (7.5)$$

being a the difference between c_1 and c_0, and b difference between l_1 and l_0. Then,
the mole is delimited with the smallest possible rectangle, resulting a new image
containing only the mole (Fig. 7.6).

The Fig. 7.7 shows that the asymmetry can happen in relation with axis y, ver-
tically, as well as in relation with axis x, horizontally. So, the both axis have to be
considered to calculate this feature.

Fig. 7.6 Image containing only the aligned mole

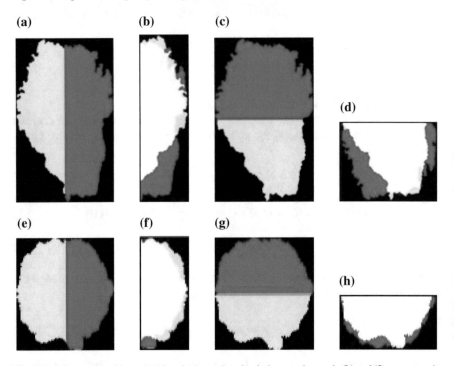

Fig. 7.7 Asymmetry with y axis (**a**) and (**e**), overlapping halves on the y axis (**b**) and (**f**), asymmetric with x axis (**c**) and (**g**) overlapping halves on the x axis (**d**) and (**h**). Melanoma (**a**), (**b**), (**c**) and (**d**) and normal mole (**e**), (**f**), (**g**) and (**h**)

It is possible identify the asymmetry difference between the melanoma, as illustrated in Figs. 7.7a–d and the normal mole, as shown in Figs. 7.7e–h by the analysis of the white region, which represents the overlapping parts of the image divided by the respective axis.

It can be observed in the non-melanotic case that the white region is occupying the largest relative area in the mole, whereas in the melanoma case this area is proportionally smaller. Lastly, we measure the mole asymmetry in percentage by the ratio between the number of pixels that do not coincide and the total number of pixels in the mole.

7.3.3   Border

To measure the irregularity of the border, the standard deviation σ_d of mole signature is calculated. Shape signature is a set of distances d_i of centroid point $pc(l_c, c_c)$, to each point $p_i(l_i, c_i)$ of the mole contours. This distance is obtained applying the Pythagorean theorem,

$$d_i = \sqrt{(l_c - l_i)^2 + (c_c - c_i)^2}. \tag{7.6}$$

Finally, the standard deviation of the signature is

$$\sigma_d = \sqrt{\sum_{i=0}^{N_d} (d_i - \mu_d)^2}, \tag{7.7}$$

being N_d the number of calculated distances and μ_d is

$$\mu_d = \frac{1}{N_d} * \sum_{i=0}^{N_d} d_i. \tag{7.8}$$

In melanoma cases this distance presents a large variance, as in Fig. 7.8a, whereas in normal moles, as in Fig. 7.8b it tend to remain constant.

Fig. 7.8 Melanoma segmented image in (**a**) and the non-cancer mole in (**b**)

(a)   **(b)**

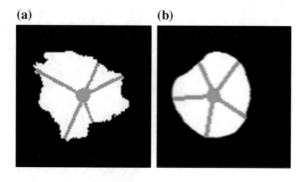

7.3.4 Color

To quantify the non-uniformity of the mole color, the variance σ^2 of the mole's histogram is calculated. The grayscale image used in this process; besides reducing the computational cost, the non-uniformity of the pixels intensity from the mole is maintained.

Firstly, it is calculated the mean intensity μ_i of the mole,

$$\mu_i = \frac{1}{L} * \sum_{i=0}^{L-1} i * p(i), \tag{7.9}$$

where i is the intensity value that can vary from 0 to $L - 1$. Being L the maximum number of the intensities which the pixel can represents; in case of grayscale image with eight bits, this value is 256. Finally, $p(i)$ is the probability for the intensity i is included in the image. Thus, the variance is calculated by

$$\sigma_i^2 = \sum_{i=0}^{L-1} (i - \mu_i)^2. \tag{7.10}$$

In order to have the color variation rate of the mole c_{rate} in the range of zero and one, it was necessary normalize the result, applying the equation

$$c_{rate} = 1 - \frac{1}{1 + \left(\frac{\sigma_i^2}{L^2}\right)}, \tag{7.11}$$

according to Gonzalez et al. [9].

Fig. 7.9 Histogram of the melanoma (**a**) and the normal mole (**b**)

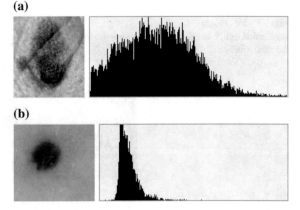

Figure 7.9a is a melanoma that presents the variance equal to 2141.9, while Fig. 7.9b has variance equal to 165.1. Therefore, it is observed that the variances of the histograms in the normal moles are usually small because the colors of the moles tend to be uniform while they are usually high in cases of melanoma.

7.4 Experimental Results

The proposed algorithms for features extraction and the models for melanoma classification are tested using images from two dermatological databases: Atlas of Clinical Dermatology and DermNet NZ. The first is a clinical dermatology atlas that has approximately 3000 images of dermatological diseases, all obtained by Niels K. Veien in his private dermatological clinic [1]. These images are intended for use in the study of dermatology area. The second, available since 1996 by New Zealand Dermatological Society Incorporated, has images and papers about skin. It is written and reviewed by health professionals and medical writers, with free access to the dataset via internet [6]. The Fig. 7.10 presents some image examples of these databases. We extract features from 139 images of moles in the skin, where 105 of these are cases of melanoma, and 34 are normal moles. All pictures are colors of 24-bit, 8-bit per channel in the RGB pattern (red, green and blue).

One way to demonstrate the antagonistic relationship between the melanoma and the normal moles for each feature is the analysis of the receiver operating characteristic (ROC) [8]. The ROC curve is a graph of true positive rates, that means the positive diagnosis with the presence of the pathology, against false positive rates, that means the negative diagnosis with the non existence of the pathology. In other words, the first rate is the ratio between the number of melanoma cases correctly classified over the total of melanoma images, while second rate is the relation between the number of normal moles misclassified as melanotic over the total cases of the normal mole

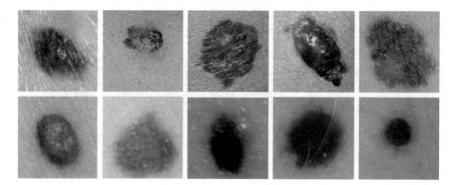

Fig. 7.10 Examples of some images of theses databases: DermNet NZ and Atlas of Clinical Dermatology

Fig. 7.11 ROC curve of the extracted features: asymmetry, border and color

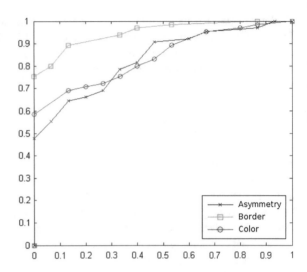

images. The quality of the result of the ROC curve is determined by the area under the curve (AUC) [4].

The Fig. 7.11 shows the comparison between the extracted features. The highest AUC obtained was 0.93 by the standard deviation of the edge of the mole. The asymmetry rate and the color rate obtained 0.82 and 0.83, respectively. So, the standard deviation of the edge can identify melanoma cases better than other features.

The results obtained using multilayer perceptron neural network (MLP), logistic regression with ridge estimator (LRR) and support vector machine (SVM) for the melanoma recognition problem are presented. We tested many different configurations for each classifier. As an evaluation approach of the models, we use the 10-fold cross-validation [7], where we divide the database into 10 sets mutually exclusive. At each iteration, one set is used for testing and the remaining sets are used for model training. The Tables 7.1, 7.2 and 7.3 present the evaluated configurations and accuracy obtained for each model, MLP, LRR and SVM, respectively. The accuracy is defined by the number of images classified correctly divided by the total number

Table 7.1 Accuracy for Melanoma classification using MLP

MLP	Configuration	Accuracy (%)
MLP-1	$n_h = 3 \mid \alpha = 0.3 \mid m = 0.2$	81.3
MLP-2	$n_h = 10 \mid \alpha = 0.3 \mid m = 0.2$	84.9
MLP-3	$n_h = 20 \mid \alpha = 0.3 \mid m = 0.2$	83.5
MLP-4	$n_h = 10 \mid \alpha = 0.2 \mid m = 0.1$	79.9
MLP-5	$n_h = 10 \mid \alpha = 0.35 \mid m = 0.2$	80.6

Table 7.2 Accuracy for Melanoma classification using logistic regression with ridge estimator

LRR	Configuration	Accuracy (%)
LRR-1	$ridge = 0.1$	84.2
LRR-2	$ridge = 0.3$	85.6
LRR-3	$ridge = 0.5$	86.3
LRR-4	$ridge = 0.8$	85.6
LRR-5	$ridge = 1$	84.9

Table 7.3 Accuracy for Melanoma classification using SVM with sigmoid kernel

SVM	Configuration	Accuracy (%)
SVM-1	$gamma = 2 \mid coef = 2 \mid cost = 1$	85.6
SVM-2	$gamma = 2.5 \mid coef = 2 \mid cost = 1$	77.0
SVM-3	$gamma = 2 \mid coef = 1 \mid cost = 1$	84.9
SVM-4	$gamma = 2 \mid coef = 1 \mid cost = 3$	84.2
SVM-5	$gamma = 2 \mid coef = 3 \mid cost = 3$	86.3

of images. The best configuration of each model was selected for a more detailed study including the MLP-2 with 84.9 % of accuracy, the LRR-3, and the SVM-5 with 86.3 % of accuracy.

The Fig. 7.12 shows the ROC curves of the best configurations for each model. The SVM-5 had the best performance with 0.867 of AUC in comparison with MLP-2, which has 0.846 of AUC and e LRR-3 with 0.851 of AUC.

The Tables 7.4, 7.5 and 7.6 represent the confusion matrix of the models. The lines correspond to the real values (target) of classes and columns correspond to the values of the output of the model (predicted). The analysis of the confusion matrix in medical diagnostic systems is important to the detriment of the comparison

Fig. 7.12 Comparison of MLP-2, LRR-3 and SVM-5 classifiers

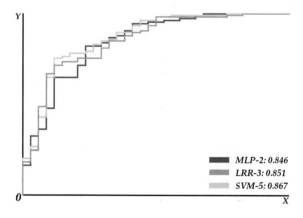

MLP-2: 0.846
LRR-3: 0.851
SVM-5: 0.867

Table 7.4 Confusion matrix
of Melanoma classification
using MLP-2

	Normal	Melanoma
Normal	18	16
Melanoma	5	100

Table 7.5 Confusion matrix
of Melanoma classification
using LRR-3

	Normal	Melanoma
Normal	17	17
Melanoma	2	103

Table 7.6 Confusion matrix
of Melanoma classification
using SVM-5

	Normal	Melanoma
Normal	19	15
Melanoma	4	101

between false positive and false negative rates. Considering the problem of skin cancer classification, we can say that the minimization of false negative rates is crucial because it represents the reduction of the error where a skin cancer was classified as a normal mole, not masking the presence of a malignancy in the patient. This type of error must be avoided since the time is a critical factor in the success of the treatment. Rather, a consider number of false positive is not considered a serious mistake, since for the patient would generate only a warning about the presence of disease.

Thus, despite not having the largest area under the curve, the model LRR-3 has the lowest number of false negative, only 2 cases. While the SVM-5 has 4 cases and the MLP-2, with the worst performance, 5 cases of cancer signs classified as benign.

7.5 Conclusion

This work introduced three algorithms for features extraction from images of moles in human skin, capable to measure the asymmetry level, the border irregularity and the non-uniformity of the color. It is important to notice that they are invariants for scale, rotation and translation, important properties in features extraction tasks.

This work also presented models to classify the melanoma. The first was the artificial neural network multilayer perceptron, the second was the logistic regression with ridge estimator, and the last was the support vector machine. All models used as input the results of the algorithms developed here. The proposed models performed well, especially the LRR-3, with 86.4 % accuracy and only 2 instances of false negative. It was observed with the experiments that the extracted features can create a good representation of classes: melanoma and normal mole.

Moreover, good rates obtained in the experiments motivate the creation a system with user iteration, for that the diameter of features and elevation are taken into account, and may improve the rates obtained in the experiments.

References

1. Atlas of Clinical Dermatology Homepage (2015). http://www.danderm-pdv.is.kkh.dk/atlas/index.html. Accessed May 2015
2. Bhuiyan AH, Azad I, Uddin K (2013) Image processing for skin cancer features extraction. Int J Sci Eng Res
3. Blum A, Zalaudek I, Argenziano G (2008) Digital image analysis for diagnosis of skin tumors. Semin Cutan Med Surg
4. Bradley AP (1997). The use of the area under the ROC curve in the evaluation of machine learning algorithms. Pattern Recogn
5. Denise MA (2004) Melanoma. nursing: patient education series
6. DermNet NZ Homepage (2015). http://www.dermnetnz.org/sitemap.html. Accessed May 2015
7. Duda RO, Hart PE, Stork DG (2000) Pattern classification, 2nd edn. Wiley-Interscience, New York
8. Fawcett T (2006) An introduction to ROC analysis. Elsevier Science Inc, Pattern Recogn. Lett
9. Gonzalez RC, Woods RE (2006) Digital image processing, 3rd edn. Prentice-Hall, Inc, Upper Saddle river
10. Horn BK (1986) Robot vision, 1st edn. McGraw-Hill Higher Education, New York
11. INCA Homepage (2015). http://www.inca.gov.br. Accessed Jun 2015
12. Manousaki AG, Manios AG, Tsompanaki EA, Evgenia I (2006). A simple digital image processing system to aid in melanoma diagnosis in an everyday melanocytic skin lesion unit. Int J Dermatol
13. OpenCV Homepage (2015). http://opencv.org. Accessed May 2015
14. Otsu N (1979) A threshold selection method from gray-level histograms. IEEE Trans Syst Man Cybern. doi:10.1109/TSMC.1979.4310076
15. Raikar A, Sangani S, Hanabaratti K (2013) Diagnosis of melanomas by check-list method. In: Fourth international conference communications and networking technologies (ICCCNT)
16. Suzuki S, Be K (1985) Topological structural analysis of digitized binary images by border following. Comput Vis Gr Image Process. doi:10.1016/0734-189X(85)90016-7

Chapter 8
Human–Machine Musical Composition in Real-Time Based on Emotions Through a Fuzzy Logic Approach

Pedro Lucas, Efraín Astudillo and Enrique Peláez

In this chapter, a method for representing human emotions is proposed in the context of musical composition, which is used to artificially generate musical melodies through fuzzy logic. A real-time prototype system, for human–machine musical compositions, was also implemented to test this approach, using the emotional intentions captured from a human musician and later used to artificially compose and perform melodies accompanying a human artist while playing the chords. The proposed method was tested with listeners in an experiment with the purpose of verifying if the musical pieces, artificially created, produced emotions in them and if those emotions matched with the emotional intentions captured from the human composer.

8.1 Introduction

Emotions are particularly intrinsic to music composition and performance despite composers might have or not considered them while composing. Vickhoff [13] argued that we do not have control over our emotions, because they are triggered involuntarily and non-consciously by nature. This raised the scientists' interest about finding ways of modeling emotions computationally to drive them in musical composition.

P. Lucas (✉) · E. Astudillo · E. Peláez
Facultad de Ingeniería en Electricidad y Computación, Centro de Tecnologías
de Información, ESPOL, Guayaquil, Ecuador
e-mail: pepaluca@espol.edu.ec

E. Astudillo
e-mail: ejastudi@espol.edu.ec

E. Peláez
e-mail: epelaez@espol.edu.ec

© Springer International Publishing Switzerland 2017   143
N. Nedjah et al. (eds.), *Designing with Computational Intelligence*,
Studies in Computational Intelligence 664,
DOI 10.1007/978-3-319-44735-3_8

Bezirganyan [2] showed that a particular melody could provoke a variety of emotions to different listeners at the same time. Vickhoff [13] also showed that an emotion could be perceived distinctly by different listeners depending on who they are and the situation involved; findings which are relevant for a system when composing melodies and pretends to provoke similar effects on listeners according to the level of emotions that a human composer would.

This work is intended to help musicians in their creative process of composing musical pieces, but with a better understanding about the role emotions play in music, through a set of synthetic intelligent partners. In contrast with previous work presented in this introduction, our approach relates the corpus of melodies with the emotions that could produce on people, considering a real-time environment.

8.2 Background

This section discusses previous work related to musical composition based on emotions, and fuzzy logic applied to music.

8.2.1 Approaches for Music Composition Based on Emotions

There have been different approaches for developing systems that analyze the emotions' content on music composition. Xiao Hu et al. [8] developed *Moody*, a system that classifies and recommends songs to users based on the mood they want to express or have in that particular moment, a solution that considers the use of support vector machines and a Naive Bayes classifier.

Strapparava et al. [11] showed that music and lyrics are able to embody deep emotions. They proposed syntactic trees for relating music and lyrics, annotated with emotions on each lyric line; in this case, support vector machines were used to classify and demonstrate that musical features and lyrics can be related emotionally.

Suiter [12] proposes a novel method using concepts of fuzzy logic to represent a set of elements and rules, considering expressiveness to trace a trajectory of musical details related to composition and establishing important points for the application of fuzzy logic over musical parameters; also, Palaniappan et al. [4] used fuzzy logic to represent musical knowledge, in which a fuzzy classifier was a component of a system for knowledge acquisition intended to Carnatic musical melodies.

Xiao Hu [7] and Wieczorkowska et al. [14] used emotions as labels for organizing, searching, and accessing musical information, whereas Misztal et al. [9] exposed a different approach by extracting emotions content from text, which then are used as inspiration for generating poems. The system proposed expresses its feelings in the form of a poem according to the affective content of the text.

8.2.2 Emotions

Finding a proper definition of emotion has been controversial and a notorious problem [3, 10]. Biologists and neurologists differ in their definition and both refer to it as a subjective quality of our present state. Emotions, according to biologists, are an important steering mechanism for animals and humans. Neurologists believe that conscious observation of emotion is specific to humans [13].

Scherer [10] claims that emotions are a reproduction of various events produced by an external or internal stimuli. Those events could be measured by taking in consideration the main following aspects.

1. Continuous changes in appraisal processes at all levels of the central nervous system,
2. Motivational changes produced by the appraisal results,
3. Patterns of facial and vocal expression as well as body movements,
4. Nature of the subjectively experienced feeling state that reflects all of these component changes.

X. Hu. [6] in his work mentions 5 fundamental generalizations of mood and their relation with music, which tell us that:

1. Mood effect in music does exist.
2. Not all moods are equally likely to be aroused by listening to music.
3. There do exist uniform mood effects among different people.
4. Not all types of moods have the same level of agreement among listeners.
5. There is some relation between listeners' judgments on mood and musical parameters such as tempo, dynamics, rhythm, timbre, articulation, pitch, mode, tone attacks, and harmony.

8.2.3 Fuzzy Logic in Music

Palaniappan et al. [4] proposed a knowledge acquisition method using a *fuzzy classifier* with the goal of representing patterns, which then could be used to generate style-based music. In this case, the notes from melodic samples are analyzed and their membership degrees are obtained by the occurrences of *established patterns* on each sample.

Suiter [12] also proposed to relate fuzzy logic principles with musical elements. In this work, the elements to represent knowledge are non-liner parameters like timbre, rhythm, frequency, and amplitude more than linear elements like notes. The focus is on expressiveness, where fuzzy sets are managed through a fuzzy controller.

In this work, fuzzy logic is used to represent *melody patterns* where emotions are denoted as fuzzy sets with membership degrees in the interval [0, 100], where 0 means absence of feelings associated to an emotion, and 100 a complete match of a feeling regarding an emotion, these values are subjective and assigned based on a human perception. Therefore, each melody pattern is labeled with emotions and, their corresponding membership degree, will represent an *emotional intention* that we will use in the fuzzification–defuzzification process.

8.3   Compositional Model Based on Emotions

The proposed model for musical composition will be used for producing melodies artificially and accompanying a human artist during a musical performance.

8.3.1   Architecture for Musical Knowledge Elicitation and Representation

The architecture for musical knowledge elicitation and representation proposed in [1], was designed based on the criteria of two experimental musicians and algorithmic compositional methods, as illustrated in Fig. 8.1, which emphasizes musical composition through a fuzzy logic approach. This architecture considers a *knowledge base* that is composed by transition matrices, obtained through a *Markov chain process* over melodies provided by human musicians and labeled with emotions by them.

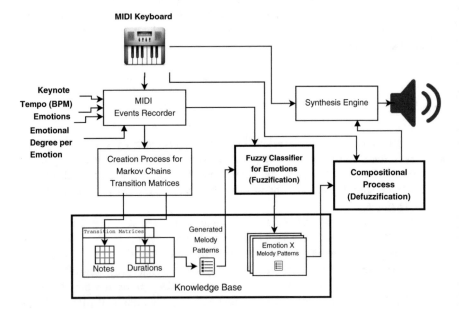

Fig. 8.1 Knowledge elicitation and representation, with compositional process

From these matrices, new melody patterns are generated and emotions are assigned through a *fuzzy classifier* implemented with a *fuzzification process* as described in Sect. 8.3.2. These patterns are then stored in the *knowledge base* for later use in the *compositional process* that produces musical pieces which are played back by a *synthesis engine* through the speakers.

Based on this architecture, a software prototype was implemented. A human musician trained the system for nurturing the *knowledge base* by performing melodies and providing the corresponding entries as depicted in Fig. 8.1. The input parameters, *emotions* and *emotional degree per emotion*, represent the *emotional intention* the musician wants to provoke to the audience; for example, happiness(80), sadness(10), and serenity(75) are three emotions weighted by the musician to express, between 0 and 100, different intentions to produce a particular emotion (0 represents no intention and 100 represents an absolute intention to produce the emotion).

The *knowledge base* is nurtured with the melody patterns generated by the composition algorithm described in [1], which is based on Markov chains. As shown in Fig. 8.1, the melody patterns generated are then labeled with the emotions and their intentions, which are defined by the musician, through a *fuzzification process* as described in Sect. 8.3.2.

For the compositional process, a real-time system was developed to support a *musical human–machine improvisation*, in which the musician (human) plays the chords for a piece that is *composed while is playing*, during this process the system (machine) produces melodies that accompany those chords by "remembering" the *musical data from chords* in the past (*notes and durations*) and stored in the *knowledge base*; also, the musician provides the *emotional intentions* previous to the composition; that is, the *emotions* that wants to produce in the audience with the *emotional degrees*, as in the training process. All this data is used to select the right melody patterns saved in the *knowledge base*, through a *defuzzification process*, as described in Sect. 8.3.2, and produce music.

8.3.2 Fuzzy Logic Approach

Considering the emotional influence by music over humans, a linguistic variable called *emotions* will be used to capture the emotions, which are fuzzy sets that the musician provides to the model, as well as the corresponding weights to represent the *emotional intentions*; therefore, the melodies provided by the musician to train the model, will create a set of solutions (melody patterns) that can be recalled later for music composition in real-time using the provided emotions.

The approach for musical composition entails two main processes. First, a fuzzification process which allows the classification of melody patterns; second a defuzzification process, which selects the piece of melody that is played back during real-time composition.

Fuzzification Process: Classification Method

The melody patterns are generated applying Markov chains as described in the architecture shown in Fig. 8.1 [1], and represented using Eq. (8.1), where *note* is an integer between 0 and 127, representing a MIDI number for the musical note, and *duration* is a relative time that is based on the *tempo* (beats per minute, BPM) which marks the rhythm for the melody pattern.

$$\textbf{Melody Pattern} : (note_0, duration_0), \qquad (8.1)$$
$$(note_1, duration_1),$$
$$\cdots, (note_n, duration_n)$$

The relative time durations ($duration_i$) that were described above, are taken from this fixed array of float numbers [0.25, 0.5, 0.75, 1, 1.25, 1.5, 1.75, 2, 2.25, 2.5, 2.75, 3, 3.25, 3.5, 3.75, 4, 4.25, 4.5, 4.75], which are representations for musical durations [5], where 1 is a *quarter note* ♩, as a reference for obtaining other durations.

The subscript n from (8.1) represents the size and it is an input parameter that controls the number of generated pairs (*note, duration*) that will compose the melody pattern. In this approach, *musical rests* [5] (intervals of silence) are not considered in the melody pattern because they are merged with their immediate previous note in order to reduce the complexity of the representation for the pattern.

The transition matrices are built from the set of melodies played by a human composer who provides the emotions and their corresponding emotional intentions, these matrices do not use these intentions in the process of generating new melody patterns, but they are used during the classification process, as a source for labeling those new patterns of size n, with the emotions given by the musician, as described below:

1. The process is applied to notes and their durations in an independent way, so a melody pattern is split in two arrays; one for *notes* and the other for *durations*. An x melody pattern generated from Markov chains will be named as MM_x (melody machine), and the corresponding arrays are $MMnotes_x$ and $MMdurations_x$
2. The melodies recorded by humans, have a size m, and will be named as MH_y (melody human) then, the two arrays will be split, following the previous procedure; hence, we have $MHnotes_y$ and $MHdurations_y$. The next steps will get the distance between MM_x and MH_y, which will be used in the emotion's labeling for MM_x. An example for these first steps is showed in Fig. 8.2.
3. To calculate the difference between each element of the $MMnotes_x$ and $MHnotes_y$ matrices, which contain MIDI notes numbers between 0 and 127, we use the following equation:

$$\Delta note_{ij} = |MHnotes_y[i] - MMnotes_x[j]| \bmod 12, \qquad (8.2)$$

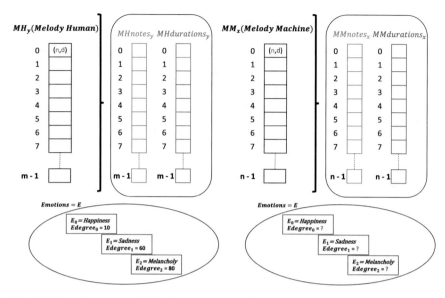

Fig. 8.2 Example corresponding to the structure used in melody patterns for human and machine

These are musical notes, which are linear musical representations composed essentially by 12 elements distributed in several octaves [5], so in this equation the octaves are not relevant because of the mod operation.

4. Equation (8.2) will be applied for each element of $MMnotes_x$ and $MHnotes_y$ to calculate the distance between $MMnotes_x$ and a segment $S_y[k]$. The segment $S_y[k]$ is a subset of size n, where k is positive integer between 0 and $m - n$, and is included in $MHnotes_y$. If $m \geq n$, there will be $m - n + 1$ segments contained in a melody created by a human, but if $m < n$, then there will be just one segment and the operations will not consider the entire $MMnotes_x$ array. Equations (8.3) and (8.4) are used to calculate the distance between $MMnotes_x$ and a segment $S_y[k]$. Figure 8.3 shows the representation of segments for $m \geq n$ and Fig. 8.4 illustrates how the distance between $MMnotes_x$ and a segment $S_y[0]$ is obtained; in this case, each $\Delta note_{ik}$ is calculated as an average, to get the required distance $ds[0]$, which in this example is 7, 00.

$$\Delta note_{ik} = |MHnotes_y[i + k] - MMnotes_x[i]| \bmod 12 , \qquad (8.3)$$

$$d(MMnotes_x, S_y[k]) = ds[k] = \frac{\displaystyle\sum_{i=0}^{\min(m,n)-1} \Delta note_{ik}}{\min(m, n)} , \qquad (8.4)$$

This model considers an average among the $\Delta note_{ik}$ values in a segment $S_y[k]$, whose results will always be a number between 0 and 11, due to the mod operation.

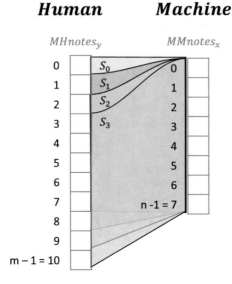

Fig. 8.3 Representation for segments in melody patterns

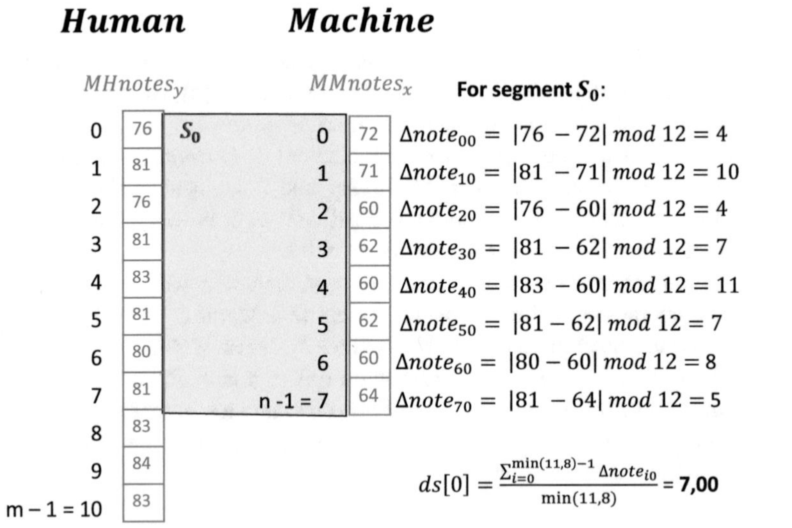

Human   Machine

	$MHnotes_y$		$MMnotes_x$		**For segment S_0:**
0	76	S_0 0	72		$\Delta note_{00} = \|76 - 72\| \bmod 12 = 4$
1	81	1	71		$\Delta note_{10} = \|81 - 71\| \bmod 12 = 10$
2	76	2	60		$\Delta note_{20} = \|76 - 60\| \bmod 12 = 4$
3	81	3	62		$\Delta note_{30} = \|81 - 62\| \bmod 12 = 7$
4	83	4	60		$\Delta note_{40} = \|83 - 60\| \bmod 12 = 11$
5	81	5	62		$\Delta note_{50} = \|81 - 62\| \bmod 12 = 7$
6	80	6	60		$\Delta note_{60} = \|80 - 60\| \bmod 12 = 8$
7	81	n -1 = 7	64		$\Delta note_{70} = \|81 - 64\| \bmod 12 = 5$
8	83				
9	84				
m − 1 = 10	83				

$$ds[0] = \frac{\sum_{i=0}^{\min(11,8)-1} \Delta note_{i0}}{\min(11,8)} = \mathbf{7{,}00}$$

Fig. 8.4 Example for calculating for distance between segment 0 and a machine melody pattern

5. The number of distances obtained for each segment, between $MMnotes_x$ and $MHnotes_y$ is $m - n + 1$; hence, to determine the minimum distance from all segments in $MHnotes_y$ and $MMnotes_x$, we use the following equation:
   For $m > n$,

$$d(MMnotes_x, MHnotes_y) = dnotes_{xy} = \min\{ds[k] : k \in [0, m - n + 1]\},$$
$$(8.5)$$

For $m < n$ there will be just one distance, that is, $d(MMnotes_x, MHnotes_y) = ds[0]$.

6. This calculation of the distance between a generated melody $MMnotes_x$ and a human melody $MHnotes_y$, has to be applied to all melodies in the *knowledge base*; therefore, if p is the number of human melodies in the *knowledge base*, then the *closest human melody*, in terms of distance and the related pattern, to $MMnotes_x$ is given by Eqs. (8.6) and (8.7).

$$DNotesMin_x = \min\{dnotes_{xy} : y \in [0, p - 1]\},\qquad (8.6)$$

$$MHnotes_{min} = MHnotes_y,\ \text{such that: } \min\{dnotes_{xy} : y \in [0, p - 1]\},$$
$$(8.7)$$

7. The human melody for notes $MHnotes_{min}$ is related with a complete melody (notes and durations) that we called $NotesMH_{min}$, which was one of the previous melodies that were labeled by a musician who established emotions and its corresponding weights along the training. This set of emotions will be denoted as E and have a size that we will call ne with a specific emotion E_r, such that r is an integer in the interval $[0, ne - 1]$. For each human melody MH, there is a set of emotions E with their corresponding weights w_r. Thus, the emotions and its weights for $NotesMH_{min}$ are used for labeling the new generated pattern $MMnotes_x$ as described in Eq. (8.8).

$$EMMnotesX_r = EMHnotesMin_r \left(1 - \frac{DNotesMin_x}{11}\right),\qquad (8.8)$$

We use 11 to normalize the minimum distance $DNotesMin_x$ for each pattern, because this value is in a range between 0 and 11, as described in step 4) of this process.

To assign weights to each emotion $EMMnotesX_r$ in the melody machine patterns $MMnotes_x$ we use Eq. (8.8) where the emotions weights, that label the *closest human melody to MMnotes_x*, come from $NotesMH_{min}$. These emotions weights are denoted as $EMHnotesMin_r$ and are weighted by $DNotesMin_x$ as the equation describes. For example, if a generated pattern ($MMnotes_x$) has a distance of 3.5 ($DNotesMin_x$) regarding its nearest human melody ($NotesMH_{min}$), and the emotions given by the musician to that melody are *happiness(10), sadness(90), and melancholy(75)* (given that $ne = 3$ and r is in $[0, 2]$), then the generated pattern will be weighted using $EMHnotesMin_r(1 - \frac{3.5}{11})$; therefore, the results are *happiness(6.82), sadness(61.36), and melancholy(51.14)*, for that generated pattern ($MMnotes_x$).

8. This process is extrapolated to *durations*; therefore, the same equations are applied to the array $MMdurs_x$, but considering: First, the difference between elements of $MMdurs_x$ and $MHdurs_y$, which will be given by Eq. (8.9).

$$\Delta dur_{ij} = \min(\left|MHdurs_y[i] - MMdurs_x[j]\right|, 4), \qquad (8.9)$$

Because durations are relative to the *tempo* (BPM), the value is fixed to 4 beats, which allows us to have a numeric reference for the normalization factor when the emotion weights are calculated. It is 4 because a *complete rhythm measure* (bar) can be basically marked as 4 beats like a metronome does [5]; and, second, the normalization factor of 4, as described before.

9. Finally, the process has to be applied to all *MM* patterns; that is, to all the generated patterns produced by the Markov chains component. Therefore, the *knowledge base* will have two kinds of weighted sets, one for notes *MMnotes* and another for the durations *MMdurs*. Since these sets are not merged in a weighted *MM*, a melody can be built with notes and durations that come from different generated patterns when the *defuzzification process* acts, providing more flexibility in the compositional process.

Defuzzification Process: Compositional Method

As in real-time musical composition (*improvisation*) [5], a *human–machine musical composition* takes place when a human musician plays the chords for a musical piece and it is accompanied by the machine or viceversa. The inputs involved are described below:

Start Input   Before initializing the system, the human musician must provide the weights for the intended emotions. These values are related to the emotional intention that the human composer wants to transmit to the audience. Also, the *tempo* (BPM) and the *keynote* have to be given.

Real-Time Input   While the musician is playing, the artificial agent gets the musical notes generated by the artist (*chords*), and produces new melodies in real time, using the data stored at the *knowledge base*.

For the compositional process, we use a metronome to guide the human composer. On every beat marked by the metronome, the system produces a melody pattern that is the result from the compositional process. This process uses the acquired notes from the previous period of time between the current beat and its predecessor, as shown in Fig. 8.5.

Not all notes of a generated melody pattern of size n are played, because of the overlapping notes in every beat. This overlapping will produce dissonance if the playing notes are still being performed in the next chord, which means that, the musician execution might not be congruent with the last set of notes [5]. To solve this problem, the melody pattern only is executed in a random number of notes between 1 and the *inputsize* (notes played by human composer). For example, if we have a melody pattern with $n = 20$ and it is received in a specific time, an input like this (48, 50, 52, 55), in MIDI notation, represents the notes (C3, D3, E3, G3), then the system will split the melody pattern in 1, 2, 3 or 4 notes. However, if the human composer plays a lot of notes, all notes will be reproduced. This behavior produces an interesting effect that makes the system generate harmonies more than melodies along the composition; an effect that is not dissonant.

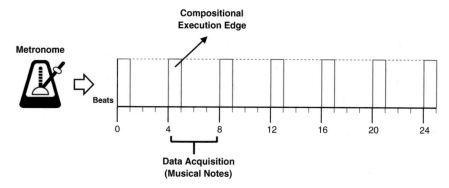

Fig. 8.5 Metronome model for data acquisition and compositional execution

To choose the melody pattern that best fits on every beat, we follow the next defuzzification procedure:

1. Since the *knowledge base* can be very wide, we need a strategy to search the best solution according to the input. Hence, the generated melody patterns are organized in *balanced binary trees*.

2. In this approach, the *knowledge base* is structured in *ne* balanced binary trees, such that *ne* is the number of emotions that are involved for all the generated patterns, where all these patterns have a certain membership degree per emotion. Thus, each tree will represent an emotion where the **keys** are the *emotional degrees* (weights or membership degrees) and the **values** are the *melody patterns* associated with that emotion. Although this representation requires some extra memory, because of the keys, it is a worthy trade-off because it helps to reduce the search space to find the right patterns and reproduce them, as we will see later.

   For example, if the *knowledge base* is trained with three emotions *happiness, sadness, and melancholy*, then we have three balanced binary trees, as illustrated in Fig. 8.6. Then, each melody pattern is added to each of those trees, based on its emotion and emotional degree; that is, if the degree of happiness is 20.0, then a new node is created in the *happiness tree* with a *key* = 20.0, and, if there is already a node with that key, then the pattern P is associated to a set of patterns which belongs to that node, in order to share this same node as in Fig. 8.7. Since the *knowledge base* have two kinds of weighted melody patterns, one for *notes* and other for *durations*, we have to consider six binary trees in this example.

3. The emotions weights provided by the musician in the *start input* might not be registered as keys in the trees, because it could not have appeared previously in the training; for example if the musician enters *happiness(10), sadness(90), and melancholy(75)*, and there is not the key 10 in the tree for happiness, then we are going to consider the *nearer keys* as shown in Fig. 8.8.

   The **nearer keys** represent the nodes that meet these requirements:

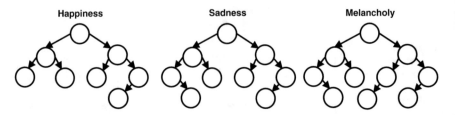

Fig. 8.6 Emotional trees examples

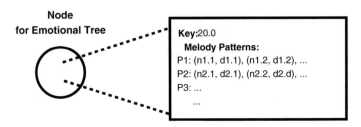

Fig. 8.7 Node structure example for an emotional tree

Fig. 8.8 Nearer keys to 10
from a happiness tree

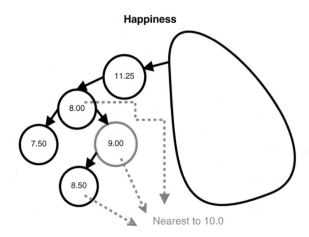

- If the target emotion weight is found, then we will collect all the melody patterns associated with this target node and its adjacent nodes; that is, the parent and its children.
- If the target is not found, we will traverse the tree until a null leaf is found, we will go up to its parent and as before, we will collect all the melody patterns from this node and from its adjacent nodes.

The objective is to have a reduced solution space with the patterns that matters for each emotion independently, which are closer to the emotion's weights given by the musician in the *start input*. This procedure can take place during the system initialization.

4. When the process is running and the human composer is playing, the system receives the notes from a MIDI keyboard (real-time input). This input and the reduced solution space trees are used for playing a melody through the speakers. To get a melody pattern, the system iterates over the reduced solution space looking for patterns whose *notes* meet the following two criteria:

- The set of weighted emotions Eh of size ne, which the human performer gave at the initialization time, is compared against all the weighted emotions Em for each pattern in the reduced space by using Eq. (8.10), which is a *Euclidean distance* to take into account all the emotions we use.

$$Dhm = \sqrt{\sum_{r=1}^{ne} (Em_r - Eh_r)^2} \, , \qquad (8.10)$$

The goal is to find the melody pattern that has the **minimum emotional distance** and also is *musically consistent* with the input notes, as explained below:

- The melody pattern to be chosen must be *musically consistent* with the harmony (chords) that the human composer is playing. Hence, we use the following criteria: *If the first note in the candidate melody pattern is part of the input provided by the human composer, then that melody pattern is consistent with the received harmony.* There could be other heuristic criteria; however,

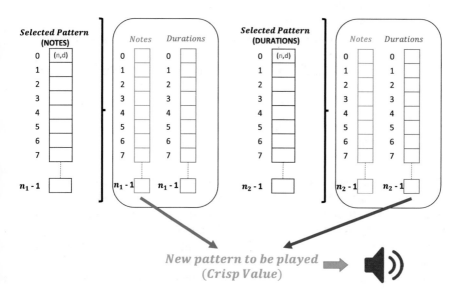

Fig. 8.9 Generation for the new melody pattern

we do not want to have a strong restriction that inhibits the artificial creativity of the system.

These two criteria are merged by an *and* ($\wedge$) operator to shape one expression and get the target melody pattern based only on *notes*. For *durations*, we just need the first criteria.

5. Finally, the two chosen arrays, *melody pattern for notes* and *melody pattern for durations* are putting together to generate a new melody. Therefore, from the fuzzy sets for *Emotions*, we get a crisp value (melody pattern) as in Fig. 8.9.

8.4 The Experiment for Musical Intention and Perceptions

8.4.1 Procedure

Musicians with academic background trained the system with 15 melodies with an average of 30.0 s per melody. The emotions selected by the artists were five: *happiness, serenity, sadness, nostalgia, and passionate*, which were weighted with emotional degrees between 0 and 100 for each melody, depending on the emotional intention, also the *keynote* and *tempo* were provided during the training. The system generated 30 melody patterns using Markov chains, which were weighted through the fuzzification process.

The human musician performed the harmonic base (chords) for 15 musical pieces approximately for 60 s, also provided the emotional intentions for each piece, such that the system played melodies that were consistent with the provided harmony and the emotional intentions using the defuzzification process.

These 15 musical pieces were then played to other people that listens western music, hence the pieces were weighted with the emotional perceptions as they were perceived, 30 people filled the assessment. A summary for this procedure is described in Fig. 8.10.

8.4.2 Results

The results presented in Fig. 8.11, as box plots, showed that for the musical piece 1, for example, the emotional intention from the musician and the perceived emotions by the listeners differed from each other; however, the perception tilts toward the emotions in a similar way as the intention. As seen in Fig. 8.11, there is more *serenity* and *nostalgia* than *happiness* and *passion* in the perceptions, as well as in the intentions, but *sadness* is not adjusted to this behavior. The plots from other musical pieces, for all emotions behaved similarly, or have one emotion that is not adjusted. Other emotions that were felt by the listeners, and not intended by the system where melancholy, reminiscence, calm, relaxation, depression, hope, and anxiety.

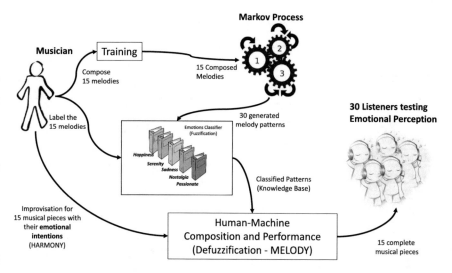

Fig. 8.10 Elements for the procedure and their interactions

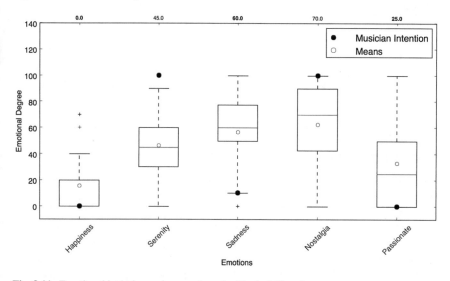

Fig. 8.11 Emotional intention and perceptions for Musical Piece 1

In Table 8.1, we present the results for a Levene's test that is used to assess the equality of variances for a variable, calculated for two or more groups. In our case, the test is applied to every song, and the groups per song are the emotions (*happiness, serenity, sadness, nostalgia, and passionate*). This test was applied with 95 % confidence and tell us that the variability for each song, regarding the emotions, does not differ significantly except for *Musical Piece 5* and *Musical Piece 6*, which means that listeners perceived each song with a similar degree of vagueness.

Table 8.1 Levene's test for the emotions' variances on each musical piece

Songs	W	P-value
Musical Piece 1	1.731	0.146
Musical Piece 2	0.377	0.824
Musical Piece 3	1.833	0.125
Musical Piece 4	1.353	0.253
Musical Piece 5	3.289	0.0129
Musical Piece 6	2.816	0.0274
Musical Piece 7	1.952	0.104
Musical Piece 8	0.753	0.557
Musical Piece 9	1.120	0.349
Musical Piece 10	0.124	0.973
Musical Piece 11	1.460	0.217
Musical Piece 12	0.968	0.427
Musical Piece 13	0.412	0.799
Musical Piece 14	0.978	0.421
Musical Piece 15	1.599	0.177

Finally, despite these results, listeners did not report any comment that suggested a random composition of melodies; though, they felt some pieces had similar melody patterns.

8.5 Conclusions

This paper presents a musical composition approach based on human emotions as fuzzy sets. The processes for fuzzification and defuzzification for these sets, were implemented in the context of a real-time system, that performed musical pieces along with human partners, who felt a well-timing execution from the artificial agent, which resulted in a proper synchronization between players, just like human musicians playing with an emotional connection. The human musicians reported that sometimes the emotional intention changed a little in order to perform consistently with the system; however, it did not affect the composition significantly, as stated by the musicians. Thus, the system is not restricted to what is required to produce, but it contributes with its own *style* to the compositional process.

This unexpected change of intentions could have caused that the emotional perception did not match significantly with the emotional intention as the results suggest. Also, this experiment did not control the emotional status for each listener, so it could have influenced in the answers; nevertheless, the variability regarding these answers is similar on each musical piece, which means that there is a subjectivity degree to be considered when people listen a song that avoids an expectation about the emotional intention. However, all the listeners reported that they felt the emotions that were established for the testing, and even other distinct emotions. These results show that the proposed method does influence on people's feelings that listen Western music.

This research contributes to the creative compositional process, providing to musicians inspirational material that is generated from the same source from which the system is trained, a style that is indeed preferred by them, based on their knowledge during the process of composing music. This approach could also be applied to other areas where real-time multimedia applications are needed; such as, video games or interactive experiences that require dynamic sound design.

Acknowledgments We thank the musicians who contributed with its criteria and talent for the basis of this work. We also thank to all people who, with patience and enthusiasm, supported us in this research as listeners.

References

1. Astudillo E, Lucas P, Peláez E (2015) Adquisición del Conocimiento en el Proceso de Composición Musical en Base a Técnicas de Inteligência Artificial. In: XI Jornadas Iberoamericanas de Ingenieria de Software e Ingenieria del Conocimiento, pp 171–185. Riobamba
2. Bezirganyan AH (2004) Analysis and estimation of emotionally colouring of music performance. In: Proceedings of the 8th international conference on music perception and cognition, pp 710–712
3. Cowie R, Sussman N, Ben-Ze'ev A (2011) Emotion: concepts and definitions
4. Ganesh Ram S, Palaniappan CT, Ramakrishnan MS, Devanathan R (2004) Knowledge engineering of creative musical expressions using Carnatic music ideology. In: International florida artificial intelligence research society conference
5. Hewitt M (2008) Music theory for computer musicians. Course Technology PTR, Boston, USA
6. Hu X (2010) Improving music mood classification using lyrics, audio and social tags. ProQuest Dissertations and Theses, p 141
7. Hu X (2010) Music and mood: Where theory and reality meet. In: Proceedings of iConference, pp 1–8. http://www.ideals.illinois.edu/handle/2142/14956
8. Hu X, Sanghvi V, Vong B (2008) Moody: a web-based music mood classification and recommendation system. In: Proceedings of the music information retrieval, pp 3–3
9. Misztal J, Indurkhya B, Poetry generation system with an emotional personality. Computationalcreativity. Net http://computationalcreativity.net/iccc2014/wp-content/uploads/2014/06/6.3_Misztal.pdf
10. Scherer KR (2005) What are emotions? And how can they be measured? doi:10.1177/0539018405058216
11. Strapparava C, Mihalcea R, Battocchi A (2007) A parallel corpus of music and lyrics annotated with emotions. In: LREC, pp 2343–2346
12. Suiter W (2010) Toward algorithmic composition of expression in music using fuzzy logic. In: Proceedings of the 2010 international conference on new interfaces for musical expression, pp 319–322
13. Vickhoff B (2008) A perspective theory of music perception and emotion, vol 90, 1st edn. Intellecta DocuSys. Västra Frölunda, Gothenburg
14. Wieczorkowska A, Synak P, Lewis R, Ras Z (2005) Extracting emotions from music data. Foundations of intelligent systems, pp 456–465. http://www.springerlink.com/index/lrc0995xxl5m12x4.pdf

Chapter 9
A Recursive Genetic Algorithm-Based Approach for Educational Timetabling Problems

Shara S.A. Alves, Saulo A.F. Oliveira and Ajalmar R. Rocha Neto

This chapter addresses the educational timetabling problem for multiple courses. This is a complex problem that basically involves a group of agents such as professors and lectures that must be weekly scheduled. The goal is to find solutions that satisfy the hard constraints and minimize the soft constraint violations. Moreover, universities often differ in terms of constraints and number of professors, courses, and resources involved, which increases the problem size and complexity. In this work, we propose a simple, scalable, and parameterized recursive approach to solve timetabling problems for multiple courses with genetic algorithms, which are efficient search methods used to achieve an optimal or near optimal solution.

9.1 Introduction

Timetabling problems are present in several enterprises and institutions. The main idea of this kind of problem is to set events into a number of time slots. Each event involves agents and may require some resources. Moreover, assignments would not violate institutions constraints that are categorized in soft and hard constraints [19]. The goal is to find solutions that satisfy the hard constraints and minimize the soft constraints violations. Educational timetabling problem is the most popular timetabling

S.S.A. Alves · S.A.F. Oliveira · A.R. Rocha Neto (✉)
Department of Teleinformatics, Federal Institute of Ceará, Fortaleza, Ceara, Brazil
e-mail: ajalmar@ifce.edu.br

S.S.A. Alves
e-mail: shara.alves@ppget.ifce.edu.br

S.A.F. Oliveira
e-mail: saulo.oliveira@ppget.ifce.edu.br

© Springer International Publishing Switzerland 2017
N. Nedjah et al. (eds.), *Designing with Computational Intelligence*,
Studies in Computational Intelligence 664,
DOI 10.1007/978-3-319-44735-3_9

problem and is a NP-hard problem, which means that the amount of computation required to find solutions increases exponentially with problem size [11]. Therefore, efficient search methods to achieve an optimal or near optimal timetable are highly desirable.

Educational timetabling problem is generally divided in three main types, to wit [19]:

1. The school timetabling that is weekly for all school lessons and avoid agents meeting two lessons at the same time;
2. The course timetabling that is weekly for all the lectures of a set of university courses and minimize overlapping lectures of courses having common students;
3. The examination timetabling for the exams of a set of university courses, spreading them for the students as much as possible and avoiding overlap of course exams having common students.

Genetic algorithms (GAs) are a search meta-heuristic method inspired by natural evolution, such as inheritance, mutation, natural selections, and crossover [12]. This meta-heuristic can be used to generate useful solutions to optimization problems. Due to the characteristics of GAs methods, it is easier to solve few kind of problems by GAs than other mathematical methods, which do have to rely on the assumption of linearity, differentiability, continuity, or convexity of the objective function [22].

Educational institutions' environments differ in terms of constraints, classes (group of students), rooms, concurrent courses, number of agents, and their unavailabilities. According to the institutions, some constraints are more important than the others. The related works often focus on one of the three main types of problem and on events involving some specific constraints, as an example the student constraints. The student constraints are commonly contemplated because students cannot attend to different events at the same time [4, 15, 21]. On the other hand, some works presuppose classes to be disjoint [1]. Nowadays, a promising constraint is the agent unavailabilities since it is common the agents work at different places [3, 8].

In this work, we proposed a simple, scalable, and parameterized model with a recursive genetic algorithm approach to solve timetabling problem for multiple courses. Such suitable model is capable to deal with different courses since universities differ from each other in number of courses and often courses lessons may change depending on the shift. The constraints are embed in fitness function and can be added or removed easily. The soft and hard constraints considered in this work were defined after an environment analysis held at Federal Institute of Ceará. The soft constraints are related to (i) adjacent lectures: more than 3 lessons in a row and (ii) agents unavailabilities: agents unavailable due other activities. The hard constraints comprehend (i) agents matches when agents are assigned to concurrent events and (ii) same course semester lectures overlapping, which minimize overlapping lectures of courses having common students. Our model solves course timetabling problems through GAs recursive executions, and as a result a global timetable is obtained.

The remaining part of this chapter is organized as follows. In Sect. 9.2, we expose some relevant related works. Then, we present our proposal in Sect. 9.3. After that, in Sect. 9.4, we describe the experiments carried out. Finally, some conclusions remarks and future works are represented in Sect. 9.5.

9.2 Related Work

A large number of approaches have been proposed for solving educational timetabling problems [18]. Most related approaches work on school and course timetabling. Spreading exams is another interesting problem to be solved, but not often the focus in most educational institutions. We found solutions with simulated annealing [1, 21], Meta-heuristic methods [7], Memetic algorithms [6, 15], genetic algorithms [2, 3, 8], and Graph-based [4, 17].

As an example, Borges's approach [3] addressed to the university timetabling problem with 8 classes and 33 agents. In such work, a mechanism for avoiding stagnation was employed, but the population size was 1000 and the time to achieve the best solution was not informed. In addition, the same timetabling problem was also covered by Ramos [8] with 14 classes, 21 agents and 10 rooms. We highlight the number of generations and runtime necessary to carry out such model, which was 15000 generations and 35 up to 90 min, respectively. We draw attention to the best fitness achieved by Ramos' model, which was 0.5 in [0,1]. Furthermore, both considered agents unavailabilities constraint.

9.3 Proposal

We aim to propose a simple, scalable, parameterized recursive genetic algorithm approach to solve timetabling problem for multiple courses. Our model suits the amount of courses, number of semesters, days and lessons per day. The global timetable is the solution, which is obtained by applying genetic algorithms (GAs) recursively so that each execution solves one course. We present our proposal from this point on.

First, the agents inform their unavailabilities in terms of period of time. Each course contains number of days and lessons per shift and a 2-tuple list [*lecture*, *agent*]. The model receives as input a list of courses, the agent unavailabilities and the GAs parameters, to wit, number of generations, mutation and crossover rate. The global timetable is achieved by recursive executions. As stated before, each GAs execution solves one course timetable and generates new assignments. These new assignments are used to update the agent unavailabilities for the next execution. Thus, after having several executions, we achieve a global timetable solution, which takes into account each and every agent unavailabilities obtained as depicted in Fig. 9.1.

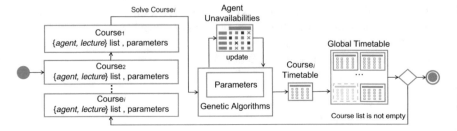

Fig. 9.1 Overview of our model

9.3.1 Genetic Algorithm

Genetic algorithms (GAs) [14] are metaheuristics in the field of artificial intelligence belonging to the larger class of evolutionary algorithms and can be used to solve optimization problems. GAs are inspired by natural evolution and mimic the process of inheritance, mutation, natural selection, and crossover. The population of species (candidate individuals or solutions) is evolved toward better individuals so that the fittest individuals remain in population. An individual has a set of chromosomes and each chromosome a set of genes, which can be changed through mutation or combined with other individual genes to generate new individuals by crossover process. The standard GA flowchart is depicted in Fig. 9.2.

First, an initial population is created and evaluated by the fitness function. Then, some selection operator is used to define which individuals are going to the reproduction step. The selected ones are crossed over and the remaining individuals may suffer mutation at random. Finally, if some stop criteria is achieved the algorithm returns its best solution or keeps evolving the population toward optimal or near optimal solution.

There are a lot of approaches, heuristics, and operators available in literature for GAs. Some main issues regarding to them require special attention, to wit, how the problem are going to be designed, named the genetic representation; which selection, crossover, and mutation operators will be chosen; and how to evaluate the individuals fitness. The aforementioned issues impact the algorithm performance in its solution search task. We present our approach design decisions in the next sections.

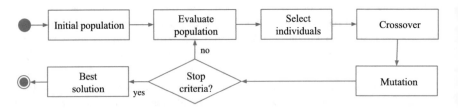

Fig. 9.2 Genetic algorithm flow chart

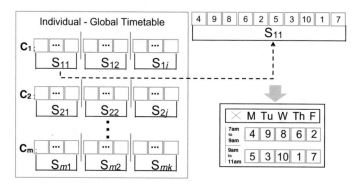

Fig. 9.3 Genetic representation

9.3.2   Genetic Representation

In Brazil, a course is divided in semesters, for example, a undergraduated course that lasts 4 years will have 8 semesters. Thus, our individual is composed by m chromosomes, which each chromosome represents a course timetable. Each timetable is divided into semesters which is a group of daily time slots as depicted in Fig. 9.3.

In our representation, we adopted integer genes to compose the chromosome and the chromosome size is defined by the number of semesters and daily time slots. Each time slot, which is a single gene, comprehends two periods of time since most lectures has two lessons at least. The gene value is unique and refers to a 2-tuple $[lecture, agent]$ or a free time in timetable.

We designed this genetic representation due its simplicity and easy verification of further restrictions, such as, agents matches. Moreover, it already avoids lectures of a same semester overlap since each gene position is related to an unique time slot.

9.3.3   Genetic Operators

Our genetic representation does not allow repeated gene values, thus a crossover operator that follows this restriction is appropriate. Among the popular crossover operators, the single point [14], multipoint [12], and uniform crossover [20] operators violate this prerogative. However, there are operators that change the order or arrangement of genes, such as OX [9], CX [16] and PMX [13]. This type of operator is called permutation operator and generates feasible individuals. In a performance comparison, it was found that OX works more effectively than the others for producing feasible course timetables [5]. We present how OX operator works in Fig. 9.4.

Nevertheless, for our genetic representation, the default OX implementation generates invalid individuals. Due to the operator to be applied to the whole individual, it would blend $[lecture, agent]$ tuples from different semesters. To overcome this

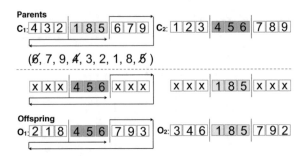

Fig. 9.4 OX crossover operator default implementation

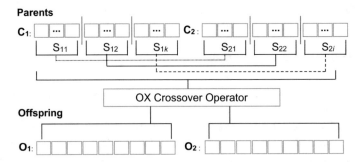

Fig. 9.5 OX crossover operator modified

Fig. 9.6 Swap mutation operator

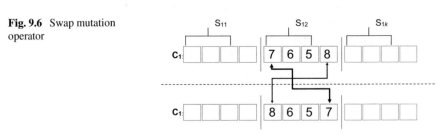

drawback, we employed a slight modification by only crossing over semester block pairs, so now it works properly. Our modification is depicted in Fig. 9.5.

As for the mutation operator, we carried out some modifications on the Swap Mutation Operator [10]. This modified operator mutates only a random semester block, swapping one gene to another as depicted in Fig. 9.6. Such behavior also avoids mixture of [*lectures*, *agent*] tuples from different semesters.

9.3.4 Fitness Function

In GAs processes, it is necessary evaluate how good an individual (possible solution) is relative to other in population. The fitness function performs an evaluation of each individual in population returning the fitness value. Depending on how fitness function was designed, higher fitness values or lower ones mean good solutions. This value is often used to choose individuals for selection and crossover stages, in other words, a wrong designed function can produce bad decisions in these stages generating worst individuals instead of better ones.

Our fitness function evaluates a course timetable based on its soft or hard constraints violations and it is defined as follows

$$F(C) = 1 - \frac{AM_C + AU_C + AL_C}{AM_{wc} + AU_{wc} + AL_{wc}}, \tag{9.1}$$

where the numerator (AM_C, AU_C and AL_C) indicates the constraints violated by C and the denominator is the worst case of each constraint (AM_{wc}, AU_{wc} and AL_{wc}). AM stands for the number of agents matches, AU for agents unavailabilities and AL for the adjacent lectures. The fitness value is $\in [0,1]$, in which values close to 1 represent good solutions.

9.3.5 Algorithm

In this subsection, we present two algorithm versions of our proposal. The first one is an iterative version, named SOLVE- ITERATIVE- TP(.) and the second version is a recursive one, called SOLVE- RECURSIVE- TP(.).

The algorithm parameters for the SOLVE- ITERATIVE- TP(.) are the list of courses (LC), the GAs parameters (GAP) and the agent unavailabilities (U). As a result, we have a global timetable solution (T). In the beginning, a course C from LC is selected and then solved by the GA. After that, we have a partial solution S and the new agent assignments. These assignments are used to update the agent unavailabilities, as well as the partial solution S is used to update the global timetable T. These steps are executed until LC being empty.

SOLVE- ITERATIVE- TP(LC, GAP, U)

LC :	list of courses
GAP :	parameters for GENETIC ALGORITHM
U :	agents unavailabilities

```
1  T ← empty ▷ the global timetable solution
2  for j ← 1 to LENGTH(LC)
3      do
4          C ← LC[j] ▷ course to be solved
5          S ← empty ▷ timetable solution for C
6          S ← GENETIC ALGORITHM(C, U, GAP)
7          U ← EXTRACT- ASSIGNMENTS(S) ∪ U
8          T ← T ∪ S ▷ adds the found solution S to T
9  return T
```

The parameters for the SOLVE- RECURSIVE- TP(.) are the list of courses (LC), the GAs parameters (GAP), the agent unavailabilities (U), and the global timetable (T). As a result, we have a global timetable solution (T) filled with the partial solutions S after each recursive execution. These recursive executions still until LC being empty.

SOLVE- RECURSIVE- TP(LC, GAP, U, T)

LC :	list of courses
GAP :	parameters for GENETIC ALGORITHM
U :	agents unavailabilities
T :	the global timetable solution

```
1  If EMPTY(LC)
2    return T
3  C ← POP(LC) ▷ course to be solved
4  S ← empty ▷ timetable solution for C
5  S ← GENETIC ALGORITHM(C, U, GAP)
6  U ← EXTRACT- ASSIGNMENTS(S) ∪ U
7  T ← SOLVE- RECURSIVE- TP(LC,GAP,U,T) ∪ S
8  return T
```

The order the courses that are selected is based on its complexity in terms of how many tuples it has. Thus, courses having more tuples are solved first.

9.4   Experiments and Discussion

To further validation of our proposal, we carried out some simulations using two different environments from Federal Institute of Ceará, as presented in Table 9.1. The environments are divided in two cases, namely, simple and complex. The difference is that in the simple case the agents unavailabilities were not informed.

9.4.1   Experiment Setup

In order to evaluate the GAs performance and find the its parameters, it was carried out 10 executions of 15 tests in each environment by changing GAs parameters. Also, we highlight that the stop criteria were: find fitness equals to 1 and stall time limit of 10 min. Furthermore, we analyze the fitness values over the generations and the final population on the more complex scenario we have, the computer science course (morning). All experiments were conducted on Core 2 Duo 2.26 GHz processor, 4 GB memory RAM on Windows 7 32-bit operational system.

Table 9.1 Environment: courses, shifts (morning, afternoon, and evening), number of semesters, agents and number of unavailabilities in simple and complex cases

Course	Shift	#Semesters	#Agents	#Simple	#Complex
Computer science	M	8	21	0	28
Computer network	M	1	5	0	8
Computer science	A	1	4	0	12
Informatic	A	4	10	0	13
Computer network	E	2	5	0	2

Table 9.2 Simple case tests results

Test	Mutation rate (%)	OX crossover rate (%)	Population size	μGenerations	# $fitness = 1$	μ Runtime (minutes)
Simple case						
1	1/12	35	75	16.5	10/10	0.2
2	1/12	35	150	15.8	10/10	0.37
4	1/12	50	75	15.9	10/10	0.19
5	1/12	50	150	15.9	10/10	0.41
7	1/40	50	75	17.2	10/10	0.2
8	1/40	50	150	14.7	10/10	0.37
10	**1/25**	**50**	**75**	**17.1**	**10/10**	**0.21**
11	1/25	50	150	15.5	10/10	0.4
12	1/25	50	600	13.1	10/10	1.32
13	1/25	60	75	16.2	10/10	0.19
14	1/25	60	150	14.8	10/10	0.37

9.4.2 Discussion

Observing the tests performed on simple case (with no unavailabilities previously informed by the agents), which some are presented in Table 9.2, all of them achieved the best solution. The average number of generations was up to only $\approx$18 and the average runtime was no more than $\approx$0.41 min, except the test numbered 9, which took $\approx$1.32 min, as we can see its average generation is the lower one $\approx$13.1 but the population size parameter is very high, 600 against the others 75 and 150. Thus, we can infer that large population is expensive.

Observing the complex case test results in Table 9.3, the agents unavailabilities previously informed increased the average generation number as well as the runtime

Table 9.3 Complex case tests results

Test	Mutation rate (%)	OX crossover rate (%)	Population size	μGenerations	# $fitness = 1$	μ Runtime (minutes)
Complex case						
1	1/12	35	75	391.1	6/10	0.66
2	1/12	35	150	161.6	7/10	1.85
3	1/12	35	600	63.4	8/10	5.87
4	1/12	50	75	389.7	7/10	2.28
5	1/12	50	150	299.7	9/10	3.25
6	1/12	50	600	83.4	5/10	7.4
7	1/40	50	75	493	5/10	1.8
8	1/40	50	150	141.6	9/10	2.66
9	1/40	50	600	68, 9	6/10	6.33
10	**1/25**	**50**	**75**	**493**	**10/10**	**3.7**
11	1/25	50	150	222.2	8/10	4.5
12	1/25	50	600	78	6/10	7
13	1/25	60	75	510.1	6/10	4
14	1/25	60	150	191.3	8/10	3.25
15	1/25	60	600	75.5	6/10	6.16

average. Furthermore, not all of 10 executions achieved the best solution in most GA parameters (see column # $fitness = 1$).

In those ones with larger populations (600), not all of their 10 executions could achieve the best solution, that is, even increasing the search space. Besides that, the average runtime was not less than 5 min for those who achieved it. Thus, we support that such parameters are expensive.

Among other tests, the test numbered 10 achieved the best solution in all executions with a runtime average of $\approx$3.7 min. Others results achieved best solution in less runtime average, but they did not succeed in most of the executions. Therefore, we support the test numbered 10 is more reliable, hence its parameters were defined to GAs.

The simple case results are presented in Table 9.2, they outperformed those ones on complex case. The achieved results took less than a half-minute in average runtime except in test numbered 10 which took 1.32. We can see through the 10th test results in Fig. 9.7 that how the runtime average increased on complex case in contrast with the simple one. The results of the 10 and 12th tests on complex case in which GA parameters differ in population size, 75 and 600, respectively, present how population sized 600 impacted the GAs performance. Indeed, the agents unavailabilities previously informed increase the problem complexity and large population is expensive.

Fig. 9.7 10 and 12th test executions average runtime

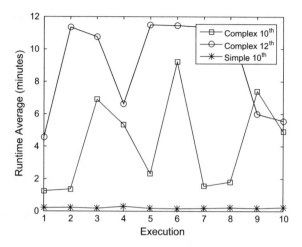

Fig. 9.8 Best fitness over the generations

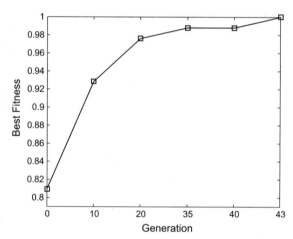

The initial population of the most complex scenario solved has already an individual with fitness value of ≈0.81 and after only 30 generations the higher fitness is increased to 0.98, see Fig. 9.8.

Furthermore, by investigating the final population, we found 5 individuals with fitness equals 1 and the average fitness of each 5 ascendant ordered individuals was ≥0.98, as shown in Fig. 9.9. Actually, our model not only solves the problem, but it also offers more than one feasible solution.

The effectiveness of our model is proved by its higher initial fitness average value and its fast convergence as presented in Table 9.4. Such model behavior is a consequence of the chosen GAs configuration and mainly due to our genetic representation, because it generates valid individuals that only violates some constraints.

It is difficult to compare our results with others due to environment complexity divergence. However, we found two similar works in which the divergences in terms

Fig. 9.9 Grouped
individuals fitness average of
43th population

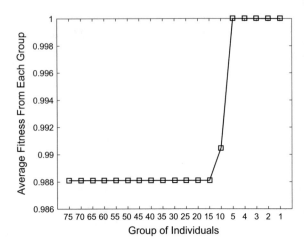

Table 9.4 Fitness average of population

Course	Generation	$\mu \, Fitness \frac{1}{3} Pop$	$\mu \, Fitness \frac{2}{3} Pop$	$\mu \, Fitness \frac{3}{3} Pop$
Computer science (Morning)	0	0.7614	0.7390	0.7136
	10	0.9195	0.9138	0.9107
	20	0.9590	0.9557	0.9536
	30	0.9790	0.9776	0.9771
	40	0.9880	0.9880	0.9880
	43	0.9909	0.9895	0.9890
Computer network (Morning)	0	0.8577	0.7977	0.7303
	1	0.96	0.9177	0.8711
Computer science (Afternoon)	0	0.875	0.836	0.7875
	2	0.9475	0.9425	0.9241
	0	0.8090	0.7799	0.7453
	10	0.9677	0.9677	0.9677
	13	0.9703	0.9690	0.9686
Computer network (Evening)	0	0.8799	0.8466	0.7955
	1	0.9500	0.9166	0.8888

of number of courses, classes, agents, and constraints are minimal [3, 8]. Both compared works solved complex and real course timetabling problem, taking into account
agents unavailabilities too. They also defined the GAs parameters after carrying out
some tests. We present such comparison in Table 9.5.

Table 9.5 Comparison with related work

Parameter	Ours	Borges' [3]	Ramos' [8]
Number of classes	16	8	14
Number of agents	21	33	21
Number of rooms	×	×	10
Population size	75	1000	100
Crossover operator	OX	PMX	Single point
Crossover rate	50 %	50 %	70 %
Mutation rate	1/25	1/20	0.002
Selection strategy	Random	Roulette wheel	Tournament
Elitism	×	10 %	×
Stagnation monitor	×	Yes	×
Generations	493	201	≈15000
Runtime (in minutes)	≈3.7	×	35–90
Fitness [0, 1]	1	1	0.5

Based on Table 9.5, we summarize the advantages and disadvantages of such works. Concerning the population size, we reinforce that larger populations are very expensive. This is confirmed through our small population size. Also, we highlight that our crossover operator helped the fast convergence against Ramos'. Usually, the number of generations is used to measure speed and robustness of discovering an acceptable solution, such aspect implies directly on the runtime. Thus, we remark those that require a higher number of generations require more runtime as well. Even our model having a number of generations exceeding Borges', the contrast of population size confirms our effectiveness.

9.5   Conclusion

This chapter presented a simple, scalable, and parameterized model to solve timetabling problems for multiple courses by applying genetic algorithms (GAs) recursively so that each execution solves one course. Universities often differ to another in number of courses and these to each other in terms of lessons per shift, for example. Such model is capable to deal with different number of courses and their features. It updates the agent unavailabilities with the new assignments after each execution.

The model was compared to similar works and the results indicate that our model took less time than the others. Also worth mentioning that the model not only repeatedly finds feasible solutions in the majority of the trials, but also finds more than one feasible solution. Future work will be aimed to enhance GAs processes, e.g., include a stagnation monitor, a heuristic for initializing population, constraints weighted, consider rooms and allow user interventions. Besides, we intend to improve our model to be capable to also solve school timetabling problems.

References

1. Abramson D (1991) Constructing school timetables using simulated annealing: sequential and parallel algorithms. Manag Sci 37(1):98–113
2. Beligiannis GN, Moschopoulos C, Likothanassis SD (2009) A genetic algorithm approach to school timetabling. J Oper Res Soc 60(1):23–42
3. Borges SK (2003) Resolução de timetabling utilizando algoritmos genéticos e evolução cooperativa. Master's thesis, Universidade Federal do Paraná
4. Burke EK, McCollum B, Meisels A, Petrovic S, Qu R (2007) A graph-based hyper-heuristic for educational timetabling problems. Eur J Oper Res 176(1):177–192
5. Chinnasri W, Krootjohn S, Sureerattanan N (2012) Performance comparison of genetic algorithm's crossover operators on university course timetabling problem. In: 2012 8th international conference on computing technology and information management (ICCM), vol 2. IEEE, New York, pp 781–786
6. Coelho AM (2006) Uma abordagem via algoritmos meméticos para a solução do problema de horário escolar. dissertação. Master's thesis, Centro Federal de Educação Tecnológica de Minas Gerais
7. de Jesus Alves RH (2010) Metaheurísticas aplicadas ao problema de horário escolar. dissertação. Master's thesis, Centro Federal de Educação Tecnológica de Minas Gerais
8. de Siqueira Ramos P (2002) Sistema automático de geração de horários para a ufla utilizando algoritmos genéticos. Master's thesis, Universidade Federal de Lavras
9. Davis L (1985) Applying adaptive algorithms to epistatic domains. IJCAI 85:162–164
10. Davis L (1991) Handbook of genetic algorithms. Van Nostrand Reinhold, New York
11. Even S, Itai A, Shamir A (1976) On the complexity of timetabling and multicommodity flow problems. SIAM J Comput 5:691–703
12. Goldberg DE (1989) Genetic algorithms in search, optimization and machine learning. Hardcover. Addison-Wesley Publishing Company, Reading
13. Goldberg DE, Lingle R (1985) Alleles, loci, and the traveling salesman problem. In: Proceedings of the international conference on genetic algorithms and their applications. Pittsburgh, PA, pp 154–159
14. Holland JH (1975) Adaptation in natural and artificial systems. MIT Press, Cambridge
15. Jat SN, Yang S (2008) A memetic algorithm for the university course timetabling problem. In: 20th IEEE international conference on tools with artificial intelligence, 2008. ICTAI'08, vol 1. IEEE, New York, pp 427–433
16. Oliver IM, Smith DJ, Holland JRC (1987) A study of permutation crossover operators on the traveling salesman problem. In: Proceedings of the second international conference on genetic algorithms on genetic algorithms and their application. L. Erlbaum Associates Inc., Hillsdale, NJ, USA, pp. 224–230. http://dl.acm.org/citation.cfm?id=42512.42542
17. Qu R, Burke EK (2009) Hybridizations within a graph-based hyper-heuristic framework for university timetabling problems. J Oper Res Soc 60(9):1273–1285
18. Qu R, Burke EK, McCollum B, Merlot LT, Lee SY (2009) A survey of search methodologies and automated system development for examination timetabling. J Sched 12(1):55–89
19. Schaerf A (1999) A survey of automated timetabling. Artif Intell Rev 13:87–127
20. Syswerda G (1989) Uniform crossover in genetic algorithms. In: Schaffer JD (ed) Proceedings of the third international conference on genetic algorithms. Morgan Kaufmann, Massachusetts, pp 2–9

21. Tuga M, Berretta R, Mendes A (2007) A hybrid simulated annealing with Kempe chain neighborhood for the university timetabling problem. In: 6th IEEE/ACIS international conference on computer and information science, 2007. ICIS 2007. IEEE, New York, pp 400–405
22. Yu L, Chen H, Wang S, Lai KK (2009) Evolving least squares support vector machines for stock market trend mining. IEEE Trans Evol Comput 13(1):87–102. doi:10.1109/TEVC.2008.928176

Chapter 10
Evolving Connection Weights of Artificial Neural Network Using a Multi-Objective Approach with Application to Class Prediction

Andrei Strickler and Aurora Pozo

In Artificial Neural Network (ANN), the selection of connection weights is a key issue and Genetic and Evolution Strategies have been found to be promising algorithms to solve this important task. Motivated by that, this study investigates the applicability of using two novel Multi-Objective Evolutionary Algorithms (MOEA): Speed constrained Multi-Objective Particle Swarm Optimization (SMPSO) and Multi-Objective Differential Evolution Algorithm based on Decomposition with Dynamical Resource Allocation (MOEA/D-DE-DRA). ANNs are training to learn data classification using sensibility and specificity for different UCI databases. The results are compared using the Hypervolume as quality indicator and statistical test.

10.1 Introduction

Most training algorithms, such as Backpropagation (BP) and conjugate gradient algorithms, are based on gradient descent [15]. There have been many successful applications of BP in various areas, but BP has drawbacks due to the use of gradient descent. It often gets trapped in a local minimum of the error function and is incapable of finding a global minimum if the error function is multimodal and/or non-differentiable.

In the other side, Evolutionary Algorithms (EAs) can help to avoid the problem of convergence to local minima and explore global search for training MLP. EAs

A. Strickler (✉) · A. Pozo
Computer Science Department, Federal University of Paraná, Curitiba, Brazil
e-mail: astrickler@inf.ufpr.br

A. Pozo
e-mail: aurora@inf.ufpr.br

© Springer International Publishing Switzerland 2017   177
N. Nedjah et al. (eds.), *Designing with Computational Intelligence*,
Studies in Computational Intelligence 664,
DOI 10.1007/978-3-319-44735-3_10

can be used effectively to find a near-optimal set of connection weights without computing gradient information. The fitness of an ANN can be defined according to different needs. Moreover, the task of learning the connection weights can be stated as a Multi-Objective task and Multi-Objective Evolutionary Algorithms (MOEAs) can be used to solve this task.

In this study, two different MOEAs are investigated: Speed constrained Multi-Objective Particle Swarm Optimization (SMPSO) [8] and Multi-Objective Differential Evolution Algorithm Based on Decomposition (MOEA/D-DE) [19] with Dynamical Resource Allocation (DRA - MOEA/D-DE-DRA) [20].

The algorithm of Speed constrained Multi-Objective Particle Swarm Optimization (SMPSO) is a technique of optimization based on Particle Swarm Optimization (PSO). PSO developed by Kennedy and Eberhart [8], is a population-based heuristic inspired by the social behavior of bird flocking aiming to find food. PSO have some similarities with evolutionary algorithms: both systems are initialized with a set of solutions, possibly random, and search for optima by updating generations. Despite these similarities, there are two main differences between them. First, there is no notion of offspring in PSO, the search is guided by the use of leaders. Secondly, PSO has no evolution operators such as crossover or mutation. In Particle Swarm Optimization, the set of possible solutions is a set of particles, called swarms moving in the search space, in a cooperative search procedure. These moves are performed by an operator that is guided by a local and a social component [9]. SMPSO algorithm is an extension of PSO for solving Multi-Objective problem. Researchers like SMPSO algorithm because this algorithm is easy to program when compared to other MOEAs.

Multi-Objective Differential Evolution Algorithm based on Decomposition (MOEA/D) is an evolutionary algorithm that optimize multi-objectives problems, using the idea of decomposition [19]. MOEA/D decompose the multi-objective problem into different sub-problems using scalar weight functions. Thus, the algorithm solves these sub problems simultaneously evolving a population of solutions using differential evolution operators. In each generation, the population is composed by the best solution found so far for each sub-problem. The relation among sub-problems are set based on the distances between their weighting vectors [19]. The MOEA/D-DE-DRA algorithm [20] uses the same concepts of MOEA/D [19], but the amount of computational resources (memory) reserved to solve each sub-problem is based on a utility function. Nowadays, MOEA/D-DRA is a state of art on MOEAs.

These two MOEAs are used to train ANN to classify data. With this purpose, two fitness functions are used: the sensitivity and specificity criteria that are directly related to the quality of the classification. An empirical evaluation is made using different UCI databases and the comparison show the effectiveness of these algorithms.

This work is structured as follow: Sect. 10.2 present the basic concepts of ANN (Sect. 10.2.1), Evolutionary Algorithms (Sect. 10.2.2), SMPSO (Sect. 10.2.2.1), MOEA/D-DE-DRA (Sect. 10.2.2.2), Hypervolume (Sect. 10.2.3) and the classification problem (Sect. 10.2.4); Sect. 10.3 describes the configuration of experiments and the obtained results. Finally, Sect. 10.4 has the conclusion and future works.

10.2 Elementary Concepts

In this section, we describe concepts of MLP, multi-objective optimization and the algorithms used in the study. Moreover, elementary concepts of classification are presented.

10.2.1 Artificial Neural Networks - ANN

Researches on neural networks look to the organization of the brain as a model for building intelligent machines. Moreover, the human brain processes information in an entirely different way than conventional digital computer [5]. The brain is a highly complex computer, non-linear and parallel. It has the ability to organize their structural components, known as neurons, in order to perform certain tasks, such as pattern recognition, sense and motor control, much faster than the fastest existing digital computer.

An ANN consists of a set of processing elements, also known as neurons or nodes, which are interconnected. It can be described as a directed graph in which each node i performs a transfer function f_i as described by Eq. 10.1

$$y_i = f_i \left(\sum_{j=1}^{n} (w_{ij} \cdot x_j) + bias \right) \tag{10.1}$$

where y_i is the output of the node i, x_i is the j_{th} input to the node, and w_{ij} is the connection weight between nodes i and j. The threshold is the *bias* of the node. Usually, f_i is nonlinear, such as a heaviside, sigmoid, or Gaussian function. Equation 10.2 shows the sigmoid function.

$$out = \frac{1}{1 + e^{-net}} \tag{10.2}$$

A neural network topology represents the way in which neurons are connected to form a network. In other words, the neural network topology can be seen as the relationship between the neurons by means of their connections. The topology of ANNs can be divided into feedforward (FFNN) and recurrent classes according to their connectivity. An ANN is a feedforward if the information flow is unidirectional. A unit sends information to another unit from which it does not receive any information. There are no feedback loops. They are used in pattern generation, recognition and classification. In recurrent ANNs, feedback loops are allowed. They are used in content addressable memories.

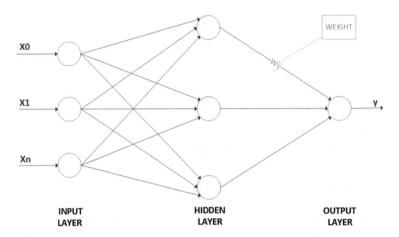

Fig. 10.1 Representation of an ANN - MLP

Basically, there are two kinds of FFNN: single-layer perceptron (SLP), and multi-layer perceptron (MLP). The SLP networks consist of a single layer of output nodes, which are fed directly by input layer via a set of weights. MLP networks consist of multiple layers: an input layer, one or more hidden layers and an output layer. Each layer has nodes and each node is fully weighted interconnected to all nodes in the subsequent layer. Figure 10.1 shows an illustration of an ANN of the type MLP.

The most important feature of an ANN is how its learning process occurs. According to Haykin [5], the learning is defined as a process where the free parameters of a neural network are adjusted by a stimulation process by the environment where it is inserted.

In supervised learning, training is performed by presenting a large set of examples, called the training set, to the network. Each example consists of a set of inputs presented to the input layer and the respective set of desired outputs. Although training an ANN can be time-consuming, once this stage is successful completed, the input–output mapping is evaluated almost instantaneously. However, care must be taken to use an adequate training set, representative of the sampling space. In many cases this is not feasible, and the sampling space must be restricted to a specific sub-domain. This means that ANNs are best applied to specific well and defined problems [3].

When using a MLP to solve a problem, the first activity is to train the MLP. Training depends on chosen initial weights and usually applies gradient learning algorithms to adapt weight values. Among these algorithms, error Backpropagation (BP) method [15] is one of the most used. In BP, the weight adjustment starts in the output nodes, where the measure of the error is available, and proceeds back-propagating this error through the previous layers. BP is a method based in gradient descendent, what means BP does not assure to find a global minimum and can get stuck on local minima, where it will stay indefinitely. However, BP is popular and widely used on ANN training [17].

As alternative, evolutionary algorithms can be applied to global searches within the weight space of a typical feedforward neural network (FFNN) and outline local minima and enable adaptive selection of control parameters [7, 16].

10.2.2 Multi-Objective Evolutionary Algorithms-MOEAs

According to Yao [18], the EAs can be used in the global evolution, to find a set of optimal (or near-optimal) weights of connections, and without gradient calculation. The error value can be defined based on the specific needs of the task to run. A commonly used factor in the formulation of the error function is the difference, called the error between the expected output and the actual output.

Two MOEAs are chosen for this study: MOEA/D-DE-DRA a state of art on MOEAs and SMPSO algorithm because this algorithm is easy to program when compared to other MOEAs.

10.2.2.1 SMPSO

Particle Swarm Optimization (PSO) is a stochastic meta-heuristic based on the movement of bird flocks looking for food, created to optimize nonlinear functions. In this method a swarm (population) of particles (solutions) moves across the search space (evolves) guided by personal and social leaders. A particle as two components: position and velocity. These components are updated at each generation.

Equations 10.3 and 10.4 present the rules for updating the speed (v_i) and position (p_i) of a particle i. The first member of the Eq. 10.3 is the inertia term, the second term is a movement to the personal best position $pBest_i^t$ and the third term is a movement towards the global best position $gBest_i^t$ (social term).

To expand the PSO to solve multi-objective problems, and create a Multi-Objective Particle Swarm Optimization (MOPSO) [14] algorithm, some modifications are needed. The first of them is the creation of an external archive (repository) to store the better (non-dominated) solutions found so far, another modification is in the leader selection scheme, which has to choose from a set of equally good leaders according to some criterion. As the number of non- dominated solutions may become very large, an archiving method is needed to prune the repository and keep only a predefined number of solutions, discarding some non-dominated solutions according to its criterion.

A MOPSO that has shown very good results in the literature is the Speed-constrained Multi-objective PSO (SMPSO) [11]. It was noted that in some conditions the velocity of the particles in a MOPSO can become too high, generating erratic movements towards the limits of the decision space. To avoid such situations, SMPSO presents a velocity constriction mechanism based on a factor χ that varies based on the values of the influence coefficients of personal and global leaders (C1 and C2 respectively). In SMPSO, the (global) leader selection method uses a binary tournament based in the Crowding Distance metric from [2], and the archiving strategy also uses the Crowing Distance.

$$v_i^{t+1} = \overbrace{\omega \cdot v_i^t}^{inertia} + \overbrace{c_1 \cdot r_1^t(pBest_i^t - p_i^t)}^{personal} + \overbrace{c_2 \cdot r_2^t(gBest_i^t - p_i^t)}^{social} \qquad (10.3)$$

$$p_i^{t+1} = p_i^t + v_i^{t+1} \qquad (10.4)$$

Algorithm 9 Pseudocode of SMPSO algorithm

Require: swarm size;
Ensure: repository;
 1: initialize(particles)
 2: repository = initializeRepository(particles)
 3: gen = 0;
 4: **while** gen < max_generations **do**
 5:   **for** each particle in the repository **do**
 6:     selectGlobalLeader(particle, repository)
 7:     ComputeSpeed(particle)
 8:     updatePosition(particle)
 9:     mutation(particle)
10:     evaluation(particle)
11:     updatePersonalLeader(particle)
12:   **end for**
13:   repository = updateRepository(particles)
14:   gen++;
15: **end while**
16: **return** repository;

At Algorithm 9 the pseudo-code of the SMPSO algorithm is presented. First the swarm and leaders archive (repository) are initialized and the evolutionary process begin. At each generation, for each particle in the population, the leaders are calculated and then the speed and position are updated. After, it is performed the Polynomial mutation for each particle, and the particles are evaluated. Finally, the particles update the leaders archive. The output of SMPSO is the leaders archive or repository.

10.2.2.2   MOEA/D-DE-DRA

The decomposition is another way to solve a problem with multi-objectives. The MOEA/D-DE-DRA decompose one multi-objective optimization problem (MOP), in many single-objective sub-problems.

There are two main components in MOEA/D. First, the mechanism to decompose MOP into sub-problems. Normally weight vectors are generated randomly and each one defines a sub-problem. The objective of each sub-problem is a (linear or nonlinear) weighted aggregation of all the individual objectives in the MOP.

The second main component is the neighborhood relations among these sub-problems. The neighborhood relations are defined based on the distances between their weight vectors. Each sub-problem (i.e., scalar aggregation function) is optimized in MOEA/D by using information from its neighboring sub-problems.

The MOEA/D-DE with Dynamical Resource Allocation (DRA) is a version where different amounts of computational effort are allocated to different problems. In MOEA/D with Dynamical Resource Allocation (MOEA/D-DE-DRA), the version of MOEA/D used in this paper, the utility π_i for each subproblem is used.

MOEA/D-DE and its variants can use any decomposition approach for defining their sub-problems. This work uses the Tchebycheff [20] approach. Using this decomposition method, each sub-problem can be formulated as in Eq. 10.5:

$$Min\ g^{te}(x \mid \lambda, z^*) = max_{1 \leq j \leq M} \{\lambda_j \mid f_j(x) - z_j^* \mid \} \tag{10.5}$$
$$subject\ to\ x \in \Omega$$

wherein g^{te} is the Tchebycheff function, $f(x) = (f_1(x), \ldots, f_M(x))$ is the set of functions that has to be minimized, and $\lambda = (\lambda_1, \ldots, \lambda_M)$ is the weight vectors.

The sub-problems are evolved using Differential Evolution(DE) operators. DE uses a simple mutation operator based on differences between pairs of solutions (called vectors) with the aim of finding a search direction based on the distribution of solutions in the current population. DE also utilizes a steady-state-like replacement mechanism, where the newly generated offspring (called trial vector) competes only against its corresponding parent (old object vector) and replaces it if the offspring has a higher fitness value.

The MOEA/D-DE-DRA is presented at Algorithm 10. The first steps of MOEA/D-DE-DRA is to initialize various data structures, analogous to most MOEA/D variants. The weight vectors λ_i, $i = 1, \ldots, N$, representing coefficients associated with each objective, are generated using a uniform distribution. The neighborhood $(B^i = i_1, \ldots, i_C)$ of weight vector λ_i stores the indexes of the C weight vectors closest to λ_i. The initial population is randomly generated and evaluated. Each individual (x_i) is associated with the i_{th} weight vector. The empirical ideal point (z^*) is initialized as the minimum value of each objective found in the initial population and the generation (g) is set to 1.

After initialization steps, the algorithm enters its main loop. The first step of the main loop is to determine which individuals from the population will be processed. A 10-tournament selection based on the utility value of each sub-problem (π_i, calculated accordingly to Eq. 10.6) is used to determine the individuals to evolve. Next, the scope used during the generation of the individual and the population update is randomly chosen. DE heuristics (mutation strategies and crossover) are applied considering individuals randomly selected from scope. In this work, scope can swap from the neighborhood to the entire population (and vice-versa) It is composed by the indexes

---

Algorithm 10 Pseudocode of MOEA/D-DE-DRA algorithm

Require: Population size (N); number of objectives (M)

 1: λ^i = genWeightVectors(N);

 2: $\lambda^i = (\lambda^i_1, ..., \lambda^i_M); i = 1, ..., N$

 3: **for** $i = 1, ..., N$ **do**

 4:    define the set of neighbor indexes $B^i = \{i_1, ..., i_C\}$, where $\{\lambda^{i_1}, ..., \lambda^{i_C}\}$ are C weight vectors closest to λ^i (Euclidian Distance)

 5: **end for**

 6: pop $\leftarrow$ initializeRandomly();

 7: Evaluate each individual $i \in$ pop and associate to its weight vector λ^i;

 8: Initialize $z^* = (z^*_1, ..., z^*_M)$;

 9: $z^*_j = min_{1 \leq i \leq N} f_j(x^i)$

10: g = 1;

11: **while** g > max evaluations **do**

12:    I = Select using 10-tournament with (π^i);

13:    **for** each Individual $i \in$ I **do**

14:      **if** *rand* $< \delta$ **then**

15:        scope = B^i;

16:      **else**

17:        scope = $\{1, ..., N\}$;

18:      **end if**

19:      y = Crossover(DE/Rand/1/bin, i);

20:      y' = PolynomialMutation(y);

21:      evaluate(y');

22:      update z^*; $z^*_j = min(z^*_j, f_j(y'))$

23:      **for** each subproblem k (k randomly selected from scope) **do**

24:        **if** $g^{te}(y'|\lambda^k, z^*) < g^{te}(x^k|\lambda^k, z^*)$ **then**

25:          **if** a new replacement may occur **then**

26:            Replace x^k by y' and increment n_r;

27:          **end if**

28:        **end if**

29:      **end for**

30:      g++;

31:    **end for**

32:    computeUtility();

33: **end while**

---

of chromosomes from either the neighborhood B^i (with probability δ) or from the entire population (with probability $1 - \delta$). Based on the chosen strategy, a modified chromosome y is generated in step 19 and modified by the polynomial mutation in step 20, generating $y' = (y'_1, \ldots, y'_n)$ from y.

In step 22, if the new chromosome y' has an objective value better than the value stored in the empirical ideal point, z^* is updated with this value. The next steps involve the population update process (steps 23–26) which is based on the comparison of the fitness of individuals. In the MOEA/D-DE framework, the fitness of an individual is measured accordingly to a decomposition function. In this work the Tchebycheff function is used (Eq. 10.5) Accordingly to what is selected for the scope (steps 15 or 17), the neighborhood or the entire population is updated.

To avoid the proliferation of y' to a great part of the population, a maximum number of updates (NR) is used. The population update is as follows: if a new replacement may occur, (i.e., while $nr < NR$ and there are unselected indexes in scope), a random index (k) from scope is chosen. If y' has a better Tchebycheff value than x_k (both using the k_{th} weight vector - λ_k) then y' replaces x_k and the number of updated chromosomes (nr) is incremented. If the current generation is a multiple of 50, then the utility value of each sub-problem is updated using Eq. 10.6. The evolutionary process stops when the maximum number of evaluations is reached.

$$\pi^2 = \begin{cases} 1, & \text{if } \Delta^i > 0.001 \\ (0.95 + 0.05 * \Delta^i/0.001) * \pi^i, & \text{otherwise} \end{cases} \qquad (10.6)$$

10.2.3   Hypervolume

The performance comparison of one or more multi-objective optimization methods is a complex task. Two goals of multi-objective optimization are: convergence and diversity of solutions.

A widely used metric in the evaluation of multi-objectives algorithms is the indicator of Hypervolume (HV). In HV, the volume of the covered area between the points of the solutions on the Pareto front P (non-dominated solutions) and a reference point W is calculated. Each solution $i \in P$, constitutes a hypercube, v_i with reference to a point W [21]. This reference point can be found by building a vector with the worst values of the objective function. The union of all hypercubes found is the result of the metric and, as higher is the value of HV better are the results. Higher values of HV indicate that there is a higher spreading between the solutions in P and indicate that there is a better convergence to the Pareto front.

Hypervolume corresponds to the area formed by the union of all rectangles, as shown in Fig. 10.2.

10.2.4   Classification Problem

Classification is one of the main tasks of Data Mining. According to Han and Kamber [4] classification is the process of finding a model or function that describes and distinguishes data elements or concepts in order to be able to use the model to predict the class of an object whose class is unknown. The derived model is based on analysis of a set of training data.

The training data consist of pairs of inputs (vectors) and desired outputs. For example, in a classification problem, a hospital may want to classify medical patients into those who have high, medium or low risk to acquiring a certain illness.

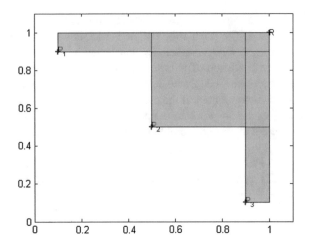

Fig. 10.2 Hypervolume area

The model generated by a learning algorithm should both fit the input data well and correctly predict the class labels of records it has never seen before. Therefore, a key objective of the learning algorithm is to build models with good generalization capability; i.e., models that accurately predict the class labels of previously unknown records.

A general approach for solving classification problems consist of two steps. First, a training set consisting of records whose class labels are known must be provided. The training set is used to build a classification model, which is subsequently applied to the test set, which consists of records with unknown class labels.

Evaluation of the performance of a classification model is based on the counts of test records correctly and incorrectly predicted by the model. These counts are tabulated in a table known as a confusion matrix (Table 10.1).

From the confusion matrix (10.1) is possible to calculate measures such as: True Positive rate (*TP rate*), True Negative rate (*TN rate* or specificity), False Positive rate (*FP rate*) and False Negative rate (*FN rate*). *TP rate*, also called sensitivity, is the precision between the positive examples (Eq. 10.7). Its complement is the *FN rate* (i.e., *FNrate* = 1 − *FPrate*). Specificity is the precision between the negative examples (Eq. 10.8). Its complement is the *FP rate*.

Table 10.1 Confusion matrix

	Class = 1	*Class* = 0	**Predicted class**
Class = 1	*TP*	*FP*	*TP* + *FP*
Class = 0	*FN*	*TN*	*FN* + *TN*
ActualClass	*TP* + *FN*	*FP* + *TN*	*N*

$$sensitivity = \frac{TP}{TP + FN} \qquad (10.7)$$

$$specificity = \frac{TN}{TN + FP} \qquad (10.8)$$

For several years, the most used performance measure for classifiers was the accuracy [1]. The accuracy is the fraction of examples correctly classified, showed on Eq. 10.9. Despite of its use, the accuracy maximization is not an appropriate goal for many of the real-world tasks [13]. A tacit assumption in the use of classification accuracy as an evaluation metric is that the class distribution among examples is constant and relatively balanced. In real world this case is rare, moreover, the cost associated with the incorrect classification of each class can be different because some classifications can lead to actions which could have serious consequences [12].

$$accuracy = \frac{TP + TN}{TP + TN + FP + FN} \qquad (10.9)$$

Classification is one of the most dynamic exploratory and application areas of ANNs. However, as mentioned before the selection of connection weights is a key issue and here this issue is tackle with two MOEAs.

10.3 Experimental Evaluation and Results

The experimental evaluation aims at answering the following research questions:

RQ1: Is there difference of performance among the configurations of each algorithm?
RQ2: Is there difference of performance between SMPSO and MOEA/D-DE-DRA?
RQ3: What are the advantages of the multi-objective versus mono-objective approach for evolving connection weights of ANN for classification task?

To answer RQ1, first different configurations of the algorithms are used to learn ANNs for each training database using sensitivity and specificity as fitness functions. Second, the learned ANNs are applied into the test databases obtaining a new set of values of sensitivity and specificity. Finally, the different configurations are compared using the Hypervolume indicator and the Friedmann rank test [6].

The goal of RQ2 is to verify whether exists one algorithm with better results than the other. The results obtained in RQ1 are now compared using the best configuration obtained for each algorithm. Again the Friedmann rank test is used.

To answer RQ3, the results generated by applying the ANNs to each test databases are analyzed using the accuracy, sensitivity and specificity.

In order to verify statistical difference among the results found by all algorithms and settings, all of them were run 30 times and Friedmann [10] and Mann–Whitney tests were executed with 0.05 significance level.

This section explains the methodology adopted to evolve connection weights of artificial neural network using a multi-objective approach and its application in class Prediction. The Java language was used to implement the ANN and to compute the two fitness functions: sensitivity and specificity. The implementation of SMPSO and MOEA/D-DE-DRA available at the JMetal Framework were used.

The following databases were used:

1. Breast Cancer Wisconsin (Original) Data Set (called as Cancer);
2. Pima Indians Diabetes Data Set (called as Diabetes);
3. Glass Identification Data Set (called as Glass);
4. Statlog (Heart) Data Set (called as Heart).

Each database was divided into 2 groups of instances, each one corresponding to training set and testing set. These groups were set up with different sizes depending on the database as shown in Table 10.2.

The topologies of the ANNs were defined according to the databases. The input layers are defined according to the numbers of attributes and the output layer according to the number of classes. The complete definition of the used topologies is presented at Table 10.3.

The topology defines the size of the individuals that were evolved by the algorithms, one dimension for each connection plus the bias for each neuron, i.e., each individual defines one ANN. The neurons used a sigmoid function.

The algorithms were executed with two different population sizes: 50 and 100 and two different number of generations: 500 and 1000, given four different configurations for each algorithm. C1 with a population size of 50 and number of generations set to 500; C2 with a population size of 50 and number of generations set to 1000; C3 with a population size of 100 and number of generations set to 500 and, C4

Table 10.2 Separation of databases

Data base	Training	Testing	Total
Cancer	500	183	683
Diabetes	650	118	768
Glass	170	44	214
Heart	220	50	270

Table 10.3 Number of neurons of each layer

Base	Attributes (Input)	Classes (Output)	Hidden
Cancer	9	2	5
Diabetes	8	2	10
Glass	9	7	10
Heart	13	2	5

Table 10.4 Parameters values used

Parameter	Value
F	0.3
CR	0.7
NR	2
T	20
Δ (delta)	0.9
c_1	[1.5:2.5]
c_2	[1.5:2.5]
r_1	[0.0:1.0]
r_2	[0.0:1.0]
ω	0.1

with a population size of 100 and number of generations set to 1000. The remaining parameters were set as presented at Table 10.4 using the default values of the JMetal.

Next we present and discuss the results of the experiments in order to answer the research questions.

10.3.1   RQ1 - Comparing Different Configuration of Each Algorithm

As mentioned before, different configurations of each algorithm were compared to set the values of the parameters: population size and number of iterations.

Table 10.5 shows the mean values and standard deviation of Hypervolume indicator. At the top of the Table, the results of the SMPSO are reported and at the bottom the results of MOEA/D-DE-DRA. For SMPSO, the best configuration for Cancer is C1, Diabetes is C4, Glass is C3 and for Heart is C4. In the case of MOEA/D-DE-DRA, the best configuration for Cancer, Diabetes and Heart is C2 and for Glass is C4. However, the difference between the values of Hypervolume is not high. For a deep analysis on these values the Kruskal–Wallis at 0.5 significance level was applied. These results are reported at Table 10.6, for SMPSO and MOEA/D-DE-DRA. Analyzing the Kruskal–Wallis results, is possible to observe that for SMPSO the configuration C4 always get best or equivalent results for all databases. For MOEA/D-DE-DRA, the configuration C2 almost always get best or equivalent results for all databases, with exception in Glass where C4 is the best configuration.

The confirmation of these findings is given by the average rankings of configurations obtained using Friedman test. These results are showed for SMPSO and MOEA/D-DE-DRA at Tables 10.7 and 10.8 respectively. Summarizing, the Friedman test point out configuration C4 for SMPSO and C2 for MOEA/D-DE-

Table 10.5 Results of Hypervolume in each configuration

Algorithm	Data Base	Mean HV C1 (Std)	Mean HV C2 (Std)	Mean HV C3 (Std)	Mean HV C4 (Std)
SMPSO	Cancer	**0.99889** (0.002572)	0.99586 (0.008267)	0.99500 (0.013317)	0.99802 (0.005244)
	Diabetes	0.81054 (0.075324)	0.85761 (0.038138)	0.85706 (0.039694)	**0.85878** (0.050186)
	Glass	0.99573 (0.001395)	0.99035 (0.003285)	**0.99681** (5.8690E-4)	0.99672 (0.001294)
	Heart	0.64645 (0.037708)	0.64999 (0.032952)	0.66789 (0.032417)	**0.67879** (0.027469)
MOEA/D-DE-DRA	Cancer	0.94478 (0.045548)	**0.97901** (0.031272)	0.97803 (0.032959)	0.94298 (0.041491)
	Diabetes	0.50173 (0.027114)	**0.62406** (0.041031)	0.60731 (0.038475)	0.50856 (0.037513)
	Glass	0.83237 (0.002012)	0.98992 (0.035803)	0.99217 (0.026678)	**0.99849** (0.002765)
	Heart	0.65033 (0.074096)	**0.72171** (0.073602)	0.71988 (0.069996)	0.69071 (0.079847)

DRA as the better considering all databases. So, these configurations were chosen for being used in the following experiments.

10.3.2 RQ2 - Comparing Different Algorithms

To answer RQ2, we compared the results from SMPSO algorithm with MOEA/D-DE-DRA, using the configurations chosen according to the results presented previously. Table 10.9 shows the results of the Wilcoxon test at 0.5 significance level and the effect size. It possible to observe that the algorithms present significant different results for each database. However, SMPSO presents better results for Cancer, Diabetes and Glass. For Heart the best results are for MOEA/D-DE-DRA.

Figure 10.3 depicts the obtained fronts using SMPSO and MOEA/D-DE-DRA for Diabetes. Tables 10.10 and 10.11 present the values of sensitivity and specificity of each of the solutions in the fronts, for SMPSO and MOEA/D-DE-DRA respectively. These fronts are the obtained fronts after executing 30 times the algorithms and removing dominated and repeated solutions.

For Diabetes, SMPSO clearly outperforms MOEA/D-DE-DRA. The same happens for Heart but, in this case, is MOEA/D-DE-DRA that outperforms SMPSO. Then, the average rankings was obtained using Friedman test. These results are presented at Table 10.12, there is possible to observe that SMPSO is slightly better than MOEA/D-DE-DRA considering the Hypervolume.

Table 10.6 Kruskal–Wallis at 0.05 significance level for Hypervolume

Dataset	Algorithm	Conf.	C1	C2	C3	C4
Cancer	SMPSO	C1	–	TRUE	TRUE	FALSE
		C2	TRUE	–	FALSE	TRUE
		C3	TRUE	FALSE	–	TRUE
		C4	FALSE	TRUE	TRUE	–
	MOEA/D-DE-DRA	C1	–	TRUE	TRUE	FALSE
		C2	TRUE	–	FALSE	TRUE
		C3	TRUE	FALSE	–	TRUE
		C4	FALSE	TRUE	TRUE	–
Diabetes	SMPSO	C1	–	TRUE	TRUE	TRUE
		C2	TRUE	–	FALSE	TRUE
		C3	TRUE	FALSE	–	TRUE
		C4	TRUE	TRUE	TRUE	–
	MOEA/D-DE-DRA	C1	–	TRUE	TRUE	FALSE
		C2	TRUE	–	FALSE	FALSE
		C3	TRUE	FALSE	–	TRUE
		C4	FALSE	FALSE	TRUE	–
Glass	SMPSO	C1	–	TRUE	TRUE	TRUE
		C2	TRUE	–	TRUE	TRUE
		C3	TRUE	TRUE	–	FALSE
		C4	TRUE	TRUE	FALSE	–
	MOEA/D-DE-DRA	C1	–	TRUE	TRUE	TRUE
		C2	TRUE	–	FALSE	TRUE
		C3	TRUE	FALSE	–	TRUE
		C4	TRUE	TRUE	TRUE	–
Heart	SMPSO	C1	–	FALSE	TRUE	TRUE
		C2	FALSE	–	FALSE	TRUE
		C3	TRUE	FALSE	–	TRUE
		C4	TRUE	TRUE	TRUE	–
	MOEA/D-DE-DRA	C1	–	TRUE	TRUE	TRUE
		C2	TRUE	–	FALSE	TRUE
		C3	TRUE	FALSE	–	TRUE
		C4	TRUE	TRUE	TRUE	–

Table 10.7 SMPSO average rankings of configurations (Friedman)

Configuration	Ranking
C1	3.0
C2	2.75
C3	2.5
C4	1.75

Table 10.8 MOEA/D-DE-DRA average rankings of configurations (Friedman)

Configuration	Ranking
C1	3.75
C2	1.5
C3	2.0
C4	2.75

Table 10.9 Wilcoxon test at 0.05 significance level, SMPSO x MOEA/D-DE-DRA, Hypervolume results

Dataset	p-value	Observation diff.	Critical diff.	Diff.	Effect size
Cancer	0.0008472	16.62222	8.837967	TRUE	0.6822841 (medium)
Diabetes	0.0009271	14.93333	8.837967	TRUE	0.7488889 (large)
Glass	0.0009148	6.81322	8.837967	TRUE	0.573216 (small)
Heart	0.0071166	12.43255	8.837967	TRUE	0.421211 (small)

10.3.3   RQ3 - Advantages of a Multi-Objective Approach

In the task of learning classification algorithms as ANNs, the goal is to create algorithms that have good performance for classification. Hence, the great majority of the methods aims to optimize the performance of the classification by improving the accuracy in the set. Despite of its use, the accuracy maximization is not an appropriate goal for many of the real-world tasks [13]. A tacit assumption in the use of classification accuracy as an evaluation metric is that the class distribution among examples is constant and relatively balanced. In real world this is rarely the case, because classification leads to actions which could have serious consequences. Therefore, recent researches point out sensitivity and specificity as better metrics to be used for induction of classification algorithms. Sensitivity is a relative measures of instances of the positive class that are well classified. Hence, the greater the

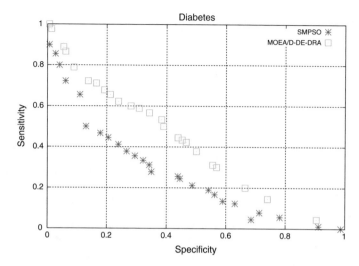

Fig. 10.3 Obtained Fronts, SMPSO and MOEAD/D-DE-DRA for Diabetes

sensitivity, the greater the number of instances in the positive class that are correctly classified. Specificity is the same measure of sensitivity, but for negative instances. The greater its value, the lower the number of instances in the negative class that are misclassified. Sensitivity and specificity are inversely proportional, meaning that as the sensitivity increases, the specificity decreases and vice versa. For understanding the advantages of a multi-objective approach based on these two metrics in the following the ANNs obtained by SMPSO for Diabetes are deeply analyzed. Table 10.13 presents TP, FP, TN, FN and accuracy of the ANNs sorted by increased value of TP. It possible to note that as TP increases, TN decreases. The best value of accuracy achieved is 0.7288135593, with TP = 127, FP = 19, TN = 45 and FN = 45. Or in other words, 127 patients that have diabetes were diagnostic as having diabetes but 19 patients that have not diabetes were included in the diagnostic. In the other hand, 45 patients that have not diabetes were confirmed as not having the diseases but 45 patients that have diabetes were diagnostic as without diabetes. This can be dangerous because a treatment at time can make a good difference on the quality of life for these patients. Having access to all these informations another ANN could be used. That is, the user as more freedom to adequate the ANN that is better for its preference.

10.4   Conclusion

ANNs are specially used to find a general solution in problems where a pattern needs to be extracted, such as data-mining. The main difficulty to apply ANN in some domain problem is to train the ANN to learn and predict. ANN provides different ways to solve many nonlinear problems that are hard to solve by conventional techniques.

Table 10.10 SMPSO obtained Pareto Front for Diabetes

Solution	Sensitivity	Specificity
s1	0.9863013699	0
s2	0.9109589041	0.0111111111
s3	0.7808219178	0.0555555556
s4	0.7123287671	0.0777777778
s5	0.6849315068	0.0444444444
s6	0.6301369863	0.1222222222
s7	0.5890410959	0.1333333333
s8	0.5616438356	0.1666666667
s9	0.5410958904	0.1888888889
s10	0.4863013699	0.2111111111
s11	0.4452054795	0.2444444444
s12	0.4383561644	0.2555555556
s13	0.3493150685	0.2777777778
s14	0.3424657534	0.3111111111
s15	0.3219178082	0.3333333333
s16	0.2945205479	0.3555555556
s17	0.2671232877	0.3777777778
s18	0.2397260274	0.4111111111
s19	0.2054794521	0.4444444444
s20	0.1780821918	0.4666666667
s21	0.1301369863	0.5
s22	0.1095890411	0.6555555556
s23	0.0616438356	0.7222222222
s24	0.0410958904	0.8
s25	0.0273972603	0.8555555556
s26	0.0068493151	0.9

The use of evolutionary algorithms has excelled to problem solving that requires space of global search optimization in several types problems. Theses algorithms have also been used to train ANNs. This paper describes and compares the results obtained in ANN training with two different algorithms: based on particle swarm optimization (SMPSO) and differential evolution(MOEA/D-DE-DRA). ANNs are trained for classification task, moreover, to properly tackle this task, ANNs need to maximize two metrics: sensitivity and specificity.

An experiment was conducted using different benchmark databases. First the goal was to determine the values of two important parameters of the algorithms: the population size and number of generations. After then, the best configurations were

Table 10.11 MOEAD obtained Pareto Front for Diabetes

Solution	Sensitivity	Specificity
S1	0.6643835616	0.2
S2	0.5684931507	0.3
S3	0.5547945205	0.3111111111
S4	0.5	0.3777777778
S5	0.4657534247	0.4222222222
S6	0.4520547945	0.4333333333
S7	0.4383561644	0.4444444444
S8	0.3904109589	0.5
S9	0.3835616438	0.5333333333
S10	0.3424657534	0.5666666667
S11	0.3082191781	0.5888888889
S12	0.2808219178	0.6
S13	0.2397260274	0.6222222222
S14	0.2123287671	0.6555555556
S15	0.1917808219	0.6777777778
S16	0.1643835616	0.7111111111
S17	0.1369863014	0.7222222222
S18	0.0890410959	0.7888888889
S19	0.0616438356	0.8666666667
S20	0.0547945205	0.8888888889
S21	0.0136986301	0.9777777778
S22	0.0068493151	1

Table 10.12 Average rankings of the algorithms (Friedman)

Algorithm	Ranking
SMPSO	2.3125
MOEAD	2.6875

compared to answer which is the best algorithm for the task. Here, it was possible to observe that the best algorithm depends on the database, however, SMPSO presented slightly better results. Finally, using the results found for Diabetes the advantages of using sensibility and specificity were illustrated.

Future works include analyzing the influence of other parameters of the algorithms, for example to use an adaptive version of MOEA/D-DE-DRA. It is known that an appropriate configuration of parameters can produce better results.

Table 10.13 SMPSO obtained solutions for Diabetes

Solution	TP	FP	TN	FN	Accuracy
S1	2	144	90	0	0.3898305085
S2	13	133	89	1	0.4322033898
S3	32	114	85	5	0.4957627119
S4	42	104	83	7	0.5296610169
S5	46	100	86	4	0.5593220339
S6	54	92	79	11	0.563559322
S7	60	86	78	12	0.5847457627
S8	64	82	75	15	0.5889830508
S9	67	79	73	17	0.593220339
S10	75	71	71	19	0.6186440678
S11	81	65	68	22	0.6313559322
S12	82	64	67	23	0.6313559322
S13	95	51	65	25	0.6779661017
S14	96	50	62	28	0.6694915254
S15	99	47	60	30	0.6737288136
S16	103	43	58	32	0.6822033898
S17	107	39	56	34	0.6906779661
S18	111	35	53	37	0.6949152542
S19	116	30	50	40	0.7033898305
S20	120	26	48	42	0.7118644068
S21	127	19	45	45	0.7288135593
S22	130	16	31	59	0.6822033898
S23	137	9	25	65	0.686440678
S24	140	6	18	72	0.6694915254
S25	142	4	13	77	0.656779661
S26	145	1	9	81	0.6525423729

Acknowledgments Authors would like to thank CNPq and CAPES for financial support.

References

1. Baronti F, Starita A (2007) Hypothesis testing with classifier systems for rule-based risk prediction (chap), pp 24–34). doi:10.1007/978-3-540-71783-6_3
2. Deb K, Agrawal S, Pratap A, Meyarivan T (2000) A fast elitist non-dominated sorting genetic algorithm for multi-objective optimisation: NSGA-II. In: Proceedings of the 6th international conference on parallel problem solving from nature, PPSN VISpringer, London, UK, pp 849–858

3. Gaspar-Cunha A, Vieira A (2005) A multi-objective evolutionary algorithm using neural networks to approximate fitness evaluations. Int J Comput Syst Signal 6(1):18–36
4. Han J, Kamber M (2006) Data mining: concepts and techniques. Morgam Kaufmann Publishers, Amsterdam
5. Haykin S (2001) Redes neurais. Bookman
6. Hodges JL, Lehmann E (2012) Rank methods for combination of independent experiments in analysis of variance. In: Rojo J (ed) Selected works of E.L. Lehmann, selected works in probability and statistics. Springer, US, pp. 403–418. doi:10.1007/978-1-4614-1412-4_35
7. Ilonen J, Kamarainen JK, Lampinen J (2003) Differential evolution training algorithm for feed-forward neural networks. Neural Process Lett 17(1):93–105. doi:10.1023/A:1022995128597
8. Kennedy J, Eberhart R (1995) Particle swarm optimization. In: Proceedings of ieee international conference on neural networks, 1995, vol 4, pp 1942–1948. doi:10.1109/ICNN.1995.488968
9. Kennedy J, Eberhart RC (2001) Swarm intelligence. Morgan Kaufmann Publishers Inc., San Francisco
10. Kruskal WH (1952) A nonparametric test for the several sample problem. Ann Math Statist 23(4):525–540. doi:10.1214/aoms/1177729332
11. Nebro AJ, Durillo JJ, Garcia-Nieto J, Coello CAC, Luna F, Alba E (2009) SMPSO: A new PSO-based metaheuristic for multi-objective optimization. In: Computational intelligence in multi-criteria decision-making, IEEE, pp. 66–73
12. Provost FJ, Fawcett T (1997) Analysis and visualization of classifier performance: comparison under imprecise class and cost distributions. In: KDD, pp 43–48
13. Provost F, Fawcett T, Kohavi R (1998) The case against accuracy estimation for comparing induction algorithms. In: proceedings 15th international conference on machine learning, Morgan Kaufmann, San Francisco, CA, pp 445–453
14. Reyes-Sierra M, Coello CAC (2006) Multi-objective particle swarm optimizers: a survey of the state-of-the-art. Int J Comput Intell Res 2(3):287–308
15. Rumelhart DE, Hinton GE, Williams RJ (1986) Learning internal representations by error propagation. Parallel distributed processing: explorations in the microstructure of cognition, vol 1. MIT Press, Cambridge, pp 318–362. http://dl.acm.org/citation.cfm?id=104279.104293
16. Slowik A (2011) Application of an adaptive differential evolution algorithm with multiple trial vectors to artificial neural network training. IEEE Trans Ind Electron 58(8):3160–3167. doi:10.1109/TIE.2010.2062474
17. van Ooyen A, Nienhuis B (1992) Improving the convergence of the back-propagation algorithm. Neural Netw 5(3):465–471. doi:10.1016/0893-6080(92)90008-7
18. Yao X (1999) Evolving artificial neural networks. Proc IEEE 87(9):1423–1447. doi:10.1109/5.784219
19. Zhang Q, Li H (2007) MOEA/D: a multiobjective evolutionary algorithm based on decomposition. IEEE Trans Evol Comput 11(6):712–731
20. Zhang Q, Liu W, Li H (2009) The performance of a new version of MOEA/D on CEC09 unconstrained MOP test instances. IEEE Congr Evol Comput 1:203–208
21. Zitzler E, Thiele L, Laumanns M, Fonseca C, da Fonseca V (2003) Performance assessment of multiobjective optimizers: an analysis and review. IEEE Trans Evol Comput 7(2):117–132. doi:10.1109/TEVC.2003.810758

Chapter 11
Diversification Strategies in Evolutionary Algorithms: Application to the Scheduling of Power Network Outages

Rainer Zanghi, Julio Cesar Stacchini de Souza and Milton Brown Do Coutto Filho

The design of evolutionary algorithms that efficiently solve complex optimization problems can be considered a challenging puzzle. In complex and multimodal problems, premature convergence to a local optimum can compromise the search for better solutions. In this work, different strategies to avoid and/or fix premature convergence of evolutionary algorithms are proposed. High diversification level is maintained throughout the evolution process, so that an adequate trade-off between solution quality and computational cost is achieved. A metric that addresses diversification in evolutionary algorithms is employed. It is shown that this metric can be used to drive the search process conveniently. The proposed diversification strategies for evolutionary algorithms are tested in a real, complex, and epistatic scheduling problem concerned with the operation of power networks. Numerical results illustrate the application of the proposed strategies and respective impact on the quality and computational cost of solutions.

R. Zanghi (✉) · M.B. Do Coutto Filho
Institute of Computing, Fluminense Federal University (UFF), Rio de Janeiro, Brazil
e-mail: rzanghi@ic.uff.br

M.B. Do Coutto Filho
e-mail: mbrown@ic.uff.br

J.C.S. de Souza
Department of Electrical Engineering and Institute of Computing,
Fluminense Federal University (UFF), Rio de Janeiro, Brazil
e-mail: julio@ic.uff.br

© Springer International Publishing Switzerland 2017
N. Nedjah et al. (eds.), *Designing with Computational Intelligence*,
Studies in Computational Intelligence 664,
DOI 10.1007/978-3-319-44735-3_11

11.1 Introduction

Since John Holland's seminal work on genetic algorithms (GAs) [18], many techniques have been proposed in the specialized literature aiming to improve the efficiency of those algorithms by enhancing the search for the optimal solution in optimization problems [13, 19]. The efficiency of the search process is usually assessed in terms of the trade-off between solution quality and computational cost. Difficult optimization problems, mainly those defined as NP-hard by the computational complexity theory [12], have encouraged researchers to propose several approaches/techniques to increase the efficiency of algorithms. As a result, for instance, GAs were encompassed by a broader terminology [19], named evolutionary algorithms (EAs).

EAs are classified as metaheuristics, which are non-exact algorithms designed to find solutions in a search space by hopefully exploring the best locations. That class of algorithms is well suited for complex optimization problems, in which global optima cannot be obtained by complete enumeration of all possible solutions, because of the high computational cost involved [3]. Therefore, the trade-off between solution quality and computational cost is an important measure to assess the performance of metaheuristics such as EAs. In addition, many real world problems have some characteristics that do not allow the use of exact optimization algorithms. Complex interdependency of variables is an example of these characteristics, being called epistasy in evolutionary computation (EC) jargon [7].

The *scheduling of network outages* (SNO) is a complex problem associated with the operation of power networks. The SNO was previously addressed as an optimization problem in [8]. The authors applied a simple elitist GA for the scheduling of optimal network outages in a small test system. However, for large networks, the computational cost of using a simple elitist GA become prohibitive, even for planning studies.

The difficulty to solve the SNO problem is mainly related to the problem instances. In addition, the behavior of the objective function is strongly dependent on the outages (to be scheduled) and power system characteristics. In this work, different strategies are proposed to reduce the overall computational cost and to make the search process more efficient, considering complex problem instances. The numerical results obtained show that the use of an EA to tackle epistasy and its adaptation to several SNO instances is promising.

The remainder of the chapter is organized in sections comprising: literature review on the application of diversification strategies; three strategies to improve the efficiency of a simple elitist GA; presentation of the SNO problem; results obtained with the application of the proposed strategies to solve the optimal SNO on IEEE benchmark networks [5]; conclusions and remarks.

11.2 Diversification Strategies and Techniques

Several authors proposed theoretical analyses on the process of an EA [13, 19, 24]. In such analyses, different optimization problems or different instances of the same problem are considered. Typically, the main objective is to assess the influence of a given strategy/technique on the quality of the obtained solution and/or the computational cost to attain it. In order to classify this influence in the search process of metaheuristics, [3] proposed a unifying view of *diversification* (ability to investigate unexplored areas of the search space) and *intensification* (the search for local improvement, in the neighborhood of the current solution). According to [3], these concepts should be considered as medium and long-term strategies based on the usage of memory acquired during the evolutionary process. The search process of an EA should balance diversification and intensification, using different techniques, in order to meet efficiency and effectiveness requirements. Operators and techniques in a simple elitist GA can be interpreted via a diversification-intensification framework. The selection operator chooses individuals for recombination based on their fitness values, favoring the intensification in the neighborhood of such individuals. The elitism operator plays a similar role, by selecting individuals with best fitness values and copying them in the next population. The crossover or recombination operator combines genotypes of two individuals previously selected in order to create new individuals. Crossover disruptive power over the genotype can be calibrated to perform intensification (small changes) or diversification (major changes). Mutation operator usually changes only one gene of the individual chromosome, which may enable the recovery of important information that was lost due to premature intensification. The adopted encoding can also interfere with the diversification-intensification capabilities of recombination and mutation operators. In Gray encoding [25], for example, the modification of more than one bit of the chromosome can have a greater impact in the decoded values than the adoption of the regular binary code, thus favoring diversification.

The concept of *genetic variation* [17], borrowed from biology, can be redefined in EC as diversification in the search process. This concept shares some key elements with its biology counterpart. The genetic variation is a metric that indicates how different genotypes are in a given population. These differences may be translated into different characteristics or traits, also called phenotype. In EC those differences may share the same properties regarding the functional relationship between genotype and phenotype. Depending on the nature of the problem, problem representation or adopted encoding, the relation between genotype and phenotype may not be biunivocal [24]. Additionally, some individuals may present the same fitness, despite of the differences between their genotypes. This situation occurs in the problem of scheduling outages addressed here; different schedules of network outages may have similar impact on the performance of the power system, being associated with the same fitness function value.

In many problems, the evaluation of the fitness function is computationally expensive, thus requiring implementation strategies that mitigate the overall computational cost of an EA. Population-based algorithms generate a large number of individuals to be assessed during the evolutionary process. Parallelization techniques—in which the evaluation of the fitness function is treated as an independent task—can reduce the overall computational cost of EAs [6]. The parallelization can occur inside the algorithm for the fitness function evaluation or explore intrinsic independence between solutions/individuals, thus placing the fitness calculation of solution candidates in different processing units. Besides, it is possible to construct a list with all different solutions already explored by the EA [24] and prevent the algorithm from evaluating several times the fitness function of the same individual. It is important to note that those techniques enable a reduction of the computational cost without affecting the solution quality. Similarly, a strategy commonly applied in discrete optimization problems is to compute and store partial fitness values [24], associated with the occurrence of specific genes, which are retrieved and used for the computation of the fitness values whenever that particular sequence of genes occurs. However, this strategy cannot be applied in epistatic problems in which a single gene cannot be directly associated with an increase or decrease in the fitness value. In this class of problems, due to dependency relations between genes, the fitness function must be completely evaluated for each unique individual [7]. Those characteristics are appealing for the use of stochastic algorithms that present high diversification [7]. However, this results in an increase of the overall computational time, as much more fitness function evaluations are necessary. In such case, the computational cost can be reduced by shortening the diversification stage of the evolutionary process (possibly compromising the solution quality) [3]. A hypothesis explored in different metaheuristics, e.g. simulated annealing, GRASP [3] and many others, is that the diversification phase should be the first one in a search process. This would allow the exploration of particular places in the search space and should be followed by an intensification phase, responsible for refining previous solutions by exploring their neighborhood. Intensification techniques should be able to make the search process converge quickly to local optima present in the neighborhood of the base solutions.

In problems that present several local optima, premature convergence can occur if the diversification phase is prematurely interrupted. This situation can be prevented in EAs with high selective pressure, when a diversification strategy may force the exploration of many different spots of the search space, possibly escaping local optima [24]. Repopulation techniques, such as the CHC algorithm proposed in [11], introduce genotypic diversity in the evolutionary process by forcing the generation of new individuals, which will be further combined. This feature is also explored by the use of *random immigrants* mechanism [16], in which a portion of the GA population is replaced by individuals that are randomly generated. In order to favor diversification but at the same time preserve the genetic material of high-quality individuals, elitism strategies are commonly used [19]. Some works in the literature report that particular encodings can be conveniently adopted to favor diversification in the evolutionary process [20, 30]. However, the benefits of many encodings, such as the Gray code [25], need further investigation.

In [1], a hierarchical distributed GA was applied to the solution of scheduling problems having epistatic characteristics similar to those found in SNO. In such problems, the importance of epistasy could be neglected and approximate fitness function evaluations were performed. However, such approach is not valid for the SNO, as epistasy can not be disregarded. Other techniques proposed to solve expensive optimization problems—by reducing the total number of fitness function evaluations or its computational cost—can be found in [27].

Many real optimization problems (e.g., the SNO) show high epistasis and adopt fitness functions whose evaluation is extremely costly. Strategies that enable the control of the diversification and intensification phases in an evolutionary process are helpful to achieve adequate algorithm performance, satisfying specific problem requirements.

11.3 Proposed Diversification Strategies

Different strategies aiming to provide high-quality solutions in a reduced computational time are presented here. Such strategies employ techniques (guided by fitness values) that promote high diversification. The proposed diversification strategies are applied to the SNO problem. It is shown that they can be employed to achieve an adequate balance between solution quality and computational efficiency. The use of a distributed fitness evaluation technique [6] also contributes to reduce the computational cost involved in the search process. This technique is also described as master–slave parallel GA in [13], distributing the fitness evaluation step of the individuals of a population among all available processor cores. This is done by using independent threads, one for the evaluation of the fitness of each individual. Further computational cost reduction is achieved by keeping a list of all unique solutions [24]. At the end of each generation, before evaluating the fitness of a given individual, its genotype is compared with those found in the list. If the individual is unique, its genotype is added to the list and its fitness evaluated. If the individual genotype is already in the list, fitness evaluation is not carried out and the individual fitness value is directly retrieved from the list. As fitness evaluation is costly in many practical problems (SNO included), this helps saving time without compromising the quality of the solutions obtained through the search process.

11.3.1 Multi-encoding

The encoding process is the transformation of phenotypic information (i.e. set of variables that constitutes an individual) into a chromosome. Figure 11.1 illustrates the SNO encoding, in which Gray, binary and integer encoding are employed to represent the situation of five different equipment out of service. In an EA, the chromosome is a genotypic representation of an individual and it allows the delimitation of a known

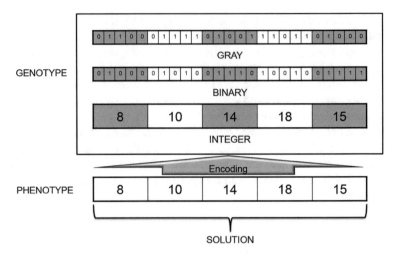

Fig. 11.1 Representation of a solution for SNO in Gray, binary and integer encodings

search space, facilitating the process of searching for an optimal solution. In [9], the encoding choice is highlighted as a key element to avoid loss of relevant information in this representation process.

The use of different encodings has been explored in the EC literature. Real encoding, as a natural representation of real or integer phenotypes, was investigated in [9]. In that work, the difficulty of propagating meaningful blocks of parents' chromosomes after crossover is discussed and a SBX crossover operator for real encoding is proposed. In [20], *delta coding* and Gray coding are considered as remapping strategies that can pose an easier search problem for a GA. It is also proposed, using delta coding, to dynamically switch the representation to avoid biases that may be associated with a particular search space representation. Reference [20] also cited the benefits of Gray encoding and its use in previous works. The reflected binary code, usually called Gray code, is a numeral system with Hamming distance always equal to one, i.e. two successive values differ in only one bit. In this mapping, it is possible to have a gradual change behavior in both genotype and phenotype, and to introduce high order changes with one-bit mutations. Reference [4] evaluates alterations in the neighborhood of an integer phenotype encoded in a Gray or binary genotype. In multimodal fitness functions, a different ordering of the neighborhood, introduced by the use of Gray encoding, can eliminate local optima that may be present in the search space when using other encodings. In [25], it is demonstrated that, in the worst case, a change to Gray encoding preserves the number of local optima present in the previous encoding.

It is important to note that changing encodings demands different mutation and recombination operators and their influence in the evolutionary process should also be considered. In [9], it is pointed out that the GA performance is affected by the choice of encoding and crossover operator pair, suggesting a harmonious combination of these two elements for the success of the optimization process. Similar recommendation is also found in [4].

In the present work, a strategy called multi-encoding is presented. This technique introduces changes in the search space representation by switching the genotype encoding during the EA evolutionary process. It explores the diversification introduced by the use of Gray encoding [31] and the computational cost reduction obtained when using integer representation. In this work, Gray encoding is first adopted and then replaced by the integer one.

Numerical results with the application of such a multi-encoding strategy to the SNO problem are also included in this chapter.

11.3.2 Repopulation with Elite Set

The *repopulation with elite set* (RES) strategy presented here is in line with the *gene flow* concept in biology. This concept [28], also called *migration*, denotes movement of genetic information from one population to the other and can be a very important source of genetic variation [17] over the migrated genes. According to Hartl and Clark, *"as an evolutionary process that brings potentially new alleles into a population, migration is qualitatively similar to mutation"*. The concept of repopulation during EA evolutionary process was addressed by delta coding algorithm [20, 30] and CHC adaptive search algorithm [11]. In delta coding, an *interim* solution obtained from the first phase of an evolutionary process is used to rescale the fitness of new individuals after a reinitialization. In CHC approach, an elite member of the population is selected to produce individuals in the next population via a specially designed operator. The previous population is removed from the evolutionary process and only new individuals generated by the operator will constitute the new population. The RES strategy presented in this chapter preserves a set of individuals from the last population, i.e., *elite set*, and populates remaining positions with new individuals. Therefore, this strategy intends to produce innovation based on interactions between a migrated elite set and a pool of new individuals. These interactions are represented by genetic operators, such as selection, recombination and mutation. In order to generate higher quality innovation from the interactions between those two pools of individuals, the RES strategy aims to produce an elite set that has individuals with a diversified genotype and good quality, measured by their fitness.

The RES algorithm executes the following two-step procedure:

- STEP 1: Candidate Selection (fitness criterion): Form a list of individuals sorted by their fitness. The set of individuals with better fitness values are selected as candidates. They are selected based on the differences between their fitnesses and the one of the best individual. The employed criterion is defined next, where p is a predefined percentage value, $FF(x_i)$ is the fitness of the i-th candidate and $FF(x^{best})$ accounts for the fitness function value obtained for the best individual.

$$FF(x_i) - FF(x^{best}) \leq p \times FF(x^{best}) \tag{11.1}$$

- STEP 2: Elite Set Formation (diversity filter): The candidate list generated at Step 1 is scanned in order to check if there are individuals genotypically too similar. If that is the case, only one of them is kept in the list. To do so, perform the following tasks:

  1. Select the next individual in the list and compare it to other candidate solutions;
  2. If the selected individual is significantly different from the other solutions, it will be moved to the elite set. Otherwise, it will be discarded. While the candidates list is not empty, return to Task 1.

The criterion used to decide whether a selected individual is similar to any other is problem-dependent. It should take into account the level of diversity to be sustained during the evolutionary process.

The RES strategy is applied through the segmentation of the evolutionary process in two nested loops: first, an internal generational loop with g generations, in which a simple elitist strategy (preserving best solution only) is employed between two generations; second, an external repopulation loop with r repopulations, in which the RES strategy is implemented. The number of individuals selected as candidates—which is dependent on the quality of the current population and the choice of the offline parameter p—can be dramatically reduced, if a diversity filter criterion is employed.

The RES technique aims to provide diversification, by introducing a pool of new individuals in the population, which replace not only low quality solutions but also good solutions that are not genotypically diverse. The same technique provides intensification (somewhat weaker) by preserving phenotypically good solutions (elitism), which may increase selection pressure.

11.3.3 Uniqueness Criterion

In [24], it is mentioned that a measure of population diversity is commonly employed to dynamically stop the evolutionary process. However, there is no consensus on diversity criteria and different methods of measuring diversity to be adopted, considering genotype, phenotype or fitness. The metrics of diversity based on genotype statistics are the most common. There are few theoretical results in this field, such as those found in [2, 10, 14, 15, 22, 29]. In [26], for example, an adaptive method based on the probability of achieving significant changes in the next generations is proposed. Many of these works have a common viewpoint, by associating GAs with stochastic search. Others apply diversity measures to guide multi-objective GAs in changing environments. Nevertheless, they all point out that threshold based methods, using statistics focused on diversification in genotypes, phenotypes or fitness values are promising strategies to guide the optimization of multimodal functions using EAs.

The *uniqueness criterion* proposed here evaluates individual novelty by its fitness value. Individuals with unique fitness values are considered new, and fitness novelty is the desired diversification characteristic to differentiate problem solutions. That criterion intends to stop the generational loop when a metric that quantifies novelty, here called *uniqueness*, falls below a given threshold value (u_t). In order to quantify the uniqueness metric, the number of new individuals generated by the EA, accumulated in a uniqueness counter u_c, is divided by the total number of individuals (tot_{ind}) created in the evolutionary process, up to that point. The uniqueness stop criterion is defined as:

$$\frac{u_c}{tot_{ind}} \leq u_t \tag{11.2}$$

By using the cardinality of unique solutions set, the uniqueness metric in (11.2) will be resistant to scaling issues usually present when statistics (e.g., average) computed in terms of the fitness values of the individuals are employed. There is no explicit dependency between the fitness value and the number of individuals with unique fitness values in a given pool. The uniqueness metric represents numerically the diversification of the evolutionary process and holds important information that can be used as a decision criterion to balance diversification and intensification. The possible sources of new individuals are the pool of individuals generated using RES and the output of genetic operations (recombination and mutation).

The uniqueness criterion adopted here is employed to calibrate automatically the number of repopulations r of the RES technique. The objective is to gain computational efficiency by reducing the number of repopulations, while keeping a high level of diversification in the evolutionary process. This approach does not require to set a maximum number of generations during the evolutionary process and the number of generations of each repopulation cycle is automatically determined. Whenever the diversification level of the evolutionary process goes below the uniqueness criterion threshold, a new repopulation cycle is triggered. Then, the number of generations in each generational loop is not fixed, as it depends on the uniqueness criterion stated in (11.2). As a result, the total number of generations in the evolutionary process tends to diminish as compared with situations in which such criterion is not employed.

11.4 Power Network Outages Scheduling

In short-term operation of power systems, it is necessary to deal with the problem of scheduling the outages of network elements, which are requested on a regular basis. In an interconnected system with different agents maintaining transmission equipment, the independent system operator (ISO) is responsible to analyze simultaneous requests and take decisions, bearing in mind that proper operating conditions, as well as system integrity, must be preserved. These analyses rely on the experience of system experts and involve considerable human and computational effort. However, only a few scheduling possibilities are covered by such analyses, which are commonly concerned only with attending operational constraints and not meeting agents requirements.

The problem of programming transmission equipment requested outages can be seen as an optimal scheduling problem, in which the resources are violations of operational constraints and deviations from the outage timetable requested by the agents. However, the problem complexity in this case is higher than that usually found in typical scheduling problems [23], owing to the interdependency between variables. The occurrence of simultaneous outages, associated with different demand profiles, can influence the network operating conditions and the representation of constraint violations. The application of exact optimization methods is impractical, due to the high computational cost to evaluate a huge number of scenarios, associated with every possible outage schedule. Metaheuristics such as EAs [19] are able to deal with both challenges (complexity and dimensionality) efficiently in complex optimization problems.

This work approaches the SNO as a constrained optimization problem, in which the objective function represents the rescheduling of a set of outages, aiming to achieve minimal deviation of the initial schedule proposed by agents, satisfying power network operational constraints. The priority of each outage, as defined by Brazilian ISO regulation [21], is also taken into account when rescheduling. Then, outages of higher priority tend not to be rescheduled or to present lower deviations from the time it was initially scheduled (as requested by the agent). Power network operational limits and the need to balance power generation and load demands are modeled as problem constraints. The optimization problem can be formulated as:

$$\min \sum_{t=1}^{n} \left(\Delta h(t) \times P_{pri}(t) \right), \qquad (11.3)$$

subject to:

$$g_k(\theta, V) = 0 \quad k = 1, \ldots, na; \qquad (11.4)$$

$$V_{ik}^{HI} - V_{ik} \geq 0 \quad i = 1, \ldots, nb; \quad k = 1, \ldots, na; \qquad (11.5)$$

$$V_{ik} - V_{ik}^{LO} \geq 0 \quad i = 1, \ldots, nb; \quad k = 1, \ldots, na; \qquad (11.6)$$

$$\left| P_{jk}^{nom} \right| - \left| P_{jk} \right| \geq 0 \quad j = 1, \ldots, nr; \quad k = 1, \ldots, na \text{ and} \qquad (11.7)$$

$$\left| P_{jk}^{emerg} \right| - \left| P_{jk} \right| \geq 0 \quad j = 1, \ldots, nr; \quad k = 1, \ldots, na. \qquad (11.8)$$

In (11.3), $\Delta h(t)$ is the deviation (hours), of the time scheduled to initiate the t-th outage, in relation to the one proposed by agents; P_{pri} is the priority associated with the t-th outage; and n is the number of outages to schedule. The equality constraint

(11.4) represents active and reactive power balance equations at each node of the network (i.e. load flow solution for the determination of complex voltages at all nodes) for each of the na scenarios to be analyzed, represented by specific topology and loading. Operational limits are represented by constraints (11.5)–(11.8), for each scenario, denoted by index k; voltage magnitudes V of nb system nodes should not violate upper and lower limits, V^{HI} and V^{LO}, respectively; and active power flows P in nr branches should not be greater than emergency limits P^{emerg} or nominal limits P^{nom}. The capability to withstand single contingencies is also taken into account, by requiring that constraints (11.5)–(11.7) be satisfied.

The evolutionary process of an EA will search for the individual with best fitness, by evaluating each proposed solution using a single fitness function, whose terms include the rescheduling deviations and established constraints, which are weighted by predefined penalty factors. Each individual of the SNO problem is represented in the EA by a chromosome where each gene contains the initial time of a given scheduled outage. The alleles of each gene are limited by the time window specified for the analyzed event occurrence. Each solution is represented as an outage schedule considering many scenarios and respective analyses, leading to high computational cost of the fitness evaluation. In SNO, this evaluation requires several runs of a power flow algorithm, which involves the solution of a set of nonlinear algebraic equations, imposing a high computational cost for large power networks and intricate schedules.

The methodology presented in [8] addressed SNO in a systematic and effective way. However, the employed optimization algorithm lacks efficiency, particularly considering its application for outage scheduling in larger power networks. In addition, the complexity and multimodality of the fitness function demands diversification strategies. The diversification strategies proposed in this chapter aim to circumvent these drawbacks, which is evidenced by the test results presented in the next section.

11.5 Simulation Results

The strategies proposed here have been incorporated in an EA to solve the SNO optimization problem. Simulations considering outages to be scheduled in the IEEE 14-, 30-, 57-, and 118-bus systems [5] were performed. For each system, the EA was executed 100 times, with initial populations randomly generated by means of different seeds. The effectiveness of the proposed approach was assessed in terms of the best solution (minimum) fitness observed in 100 executions of the EA, as well as the average of the best solutions found and the associated standard deviation. The computational efficiency was assessed in terms of the average execution time, which represents the average of the execution times observed. The EA parameters adopted were the same used in [8, 31].

11.5.1 Repopulation with Elite Set

Table 11.1 shows the results obtained for each test system, using Gray and integer encodings, with and without the RES strategy. A percentage p set to 40 % was used in (11.1) when selecting candidates to form the elite set. Besides, two individuals were considered significantly different when (in the corresponding schedules) the scheduled time of any given outage differs at least in three hours. Each execution of the EA consisted of 70 generational loops (repopulations), each one with 10 generations. When the RES strategy was not used, a single generational loop with 700 generations was considered in the evolutionary process.

The results in Table 11.1 show that the best fitness values are improved or preserved when using the RES strategy. The RES strategy also improved the average fitness values in all cases, meaning that it is more likely to obtain better solutions when such strategy is employed. In addition to the solution quality improvement enabled by the use of the RES technique, the results in Table 11.1 also reveal that better solutions are obtained by Gray than integer encoding. On the other hand, a faster convergence is achieved with integer encoding. The use of the RES strategy and Gray encoding favor diversity and allow the exploration of more unique and diverse individuals during the evolutionary process. When the RES strategy and the Gray encoding are not used, more individuals that are identical appear. The fitness functions of the same individual do not need to be recalculated, resulting in computational gain. However,

Table 11.1 Repopulation with elite set and encodings

Test case	Enc.	RES	Fitness			Time (s)
			Min.	Avg.	Std. dev.	Avg.
IEEE14	Gray	No	42.064	42.590	0.840	8
	Gray	Yes	42.064	42.354	0.576	22
	int.	No	42.064	46.108	2.647	1
	int.	Yes	42.064	42.933	0.788	14
IEEE30	Gray	No	8.449	10.648	1.424	77
	Gray	Yes	8.449	10.432	1.263	108
	int.	No	8.748	18.269	4.667	5
	int.	Yes	8.720	13.712	2.604	60
IEEE57	Gray	No	17.029	18.610	0.860	381
	Gray	Yes	17.029	18.382	0.865	508
	int.	No	17.564	23.464	4.109	16
	int.	Yes	17.029	20.317	1.553	322
IEEE118	Gray	No	10.021	17.091	4.818	1777
	Gray	Yes	10.021	16.822	4.505	3239
	int.	No	19.714	33.600	9.028	87
	int.	Yes	19.714	28.054	4.793	2126

it is important to stress that the solution quality brought by the use of the Gray encoding and RES technique is positively affected.

11.5.2 Multi-encoding

Bearing in mind the benefits brought to the quality of the obtained solutions and to the computational efficiency when using different encodings, a combination of Gray and integer encodings is proposed. Therefore, Gray encoding, which naturally favor diversification, was employed at the beginning of evolutionary process. In a later stage, aiming to accelerate the convergence of the EA, that encoding was replaced by the integer one. In all simulations, the Gray encoding was switched to the integer one after performing 50 % of the total number of generational loops of the evolutionary process. As in the previous simulations, 70 generational loops, each one consisting of 10 generations, were considered. The RES strategy was adopted in all cases.

The results obtained using the multi-encoding approach are shown in Table 11.2. For the sake of comparison, results obtained with the use the RES strategy, but using only Gray or integer encoding, extracted from Table 11.1, are also shown.

The results show that the use of the multi-encoding strategy allows a significant reduction in the computational time when compared to the use of the Gray encoding only, being negligible the impact on the quality of the obtained solutions. This shows that the multi-encoding approach can provide better trade-offs between solution quality and computational efficiency, if the diversification and intensification provided by each encoding is conveniently explored during the evolutionary process.

Table 11.2 Multi-encoding 50 %

Test case	First encod.	Sec. encod.	Fitness			Time (s)
			Min.	Avg.	Std. Dev.	Avg.
IEEE14	Gray	–	42.064	42.354	0.576	22
	int.	–	42.064	42.933	0.788	14
	Gray	int.	42.064	42.430	0.645	18
IEEE30	Gray	–	8.449	10.432	1.263	108
	int.	–	8.720	13.712	2.604	60
	Gray	int.	8.449	10.552	1.456	87
IEEE57	Gray	–	17.029	18.382	0.865	508
	int.	–	17.029	20.317	1.553	322
	Gray	int.	17.029	18.470	0.856	418
IEEE118	Gray	–	10.021	16.822	4.505	3239
	int.	–	19.714	28.054	4.793	2126
	Gray	int.	10.021	16.756	4.596	2852

11.5.3 Uniqueness Criterion

The uniqueness criterion was also explored to enhance the computational efficiency of the EA by monitoring the diversification and intensification stages during the evolutionary process. This was accomplished by observing the uniqueness criterion in order to decide when to interrupt each generational loop and repopulate again using the RES strategy. Then, whenever the *uniqueness* defined in (11.2) falls below the threshold value u_t, the generational loop is interrupted. Considering the evolutionary process as a whole, computational time savings are obtained whenever the generational loop is interrupted before the predefined number of generations (equals to 10) is achieved. It is important to note that more emphasis is given to diversification when such uniqueness criterion is adopted, as it tends to preserve more diverse solutions throughout the evolutionary process. Although in a less degree, intensification is also achieved, as the diverse elite solutions that are preserved by the RES strategy tend to be continuously enhanced after each generational loop. Since a high diversity level is sustained until the end of the evolutionary process, a simple local search (LS) is executed only once, at the end of the evolutionary process and having the best solution as the starting point, thus favoring local search intensification. The employed LS is a greedy algorithm in which new solutions are explored by changing the initial hour of each outage. Whenever a better solution is found, it becomes the base for a new local search cycle. The search stops when it is not possible to enhance the current base solution after exploring its neighborhood.

Table 11.3 presents simulation results obtained when using the uniqueness criterion. In all the simulations, the Gray encoding was adopted. As previously discussed, the number of generations in each loop was not fixed and depended on the uniqueness criteria stated in (11.2). In the performed simulations the threshold u_t was set to 0.7.

Table 11.3 Uniqueness criterion with RES

Test case	RES	Uniq. Crit.	Fitness			Time (s)
			Min.	Avg.	Std. dev.	Avg.
IEEE14	No	No	42.064	42.590	0.840	8
	Yes	No	42.064	42.352	0.574	22
	Yes	70%	42.064	42.641	0.649	4
IEEE30	No	No	8.449	10.433	1.263	77
	Yes	No	8.449	10.184	1.191	110
	Yes	70%	8.449	10.769	1.780	20
IEEE57	No	No	17.029	18.152	1.059	387
	Yes	No	17.029	18.030	0.964	512
	Yes	70%	17.029	18.525	1.083	110
IEEE118	No	No	10.021	16.840	4.652	1791
	Yes	No	10.021	16.605	4.403	3253
	Yes	70%	10.021	16.945	4.679	703

Table 11.4 Fitness values—proposed (*ideal*) versus optimized

Test case	Solution fitness	
	Proposed	Found
IEEE14	233.013	42.064
IEEE30	161.374	8.449
IEEE57	930.262	17.029
IEEE118	969.289	10.021

In Table 11.3 the results obtained using the uniqueness criterion can be compared with those in which the number of generations per generational cycle was fixed and predefined. For the sake of comparison, the Gray encoding was adopted in all cases and the same LS technique was applied at the end of evolutionary process.

The results in Table 11.3 show that considerable gains in computational efficiency are achieved when using the uniqueness criterion. The effect on the quality of the final solution can be considered negligible, as shown by the minimum and average fitness function values observed for each test system.

11.5.4 Effectiveness of the Optimization Process

As previously discussed, the optimal solution for the formulated SNO optimization problem will be the schedule that does not lead to violations of power network operational constraints, with minimal deviations from the *ideal* schedule—the one requested by network agents. Table 11.4 shows, for each test system, the fitness values associated with the solution that represents the *ideal* schedule and the best solution found by the EA (the one incorporating the strategies proposed here). Thus the fitness values associated with the *ideal* schedules are only concerned with violations of network operational constraints. It can be seen from Table 11.4 that the fitness values were substantially reduced during the evolutionary process, reflecting the optimization carried out by the EA.

11.6 Conclusions

This chapter presented different strategies to explore the diversification and intensification stages in an EA so that adequate trade-offs between solution quality and computational efficiency can be achieved when solving complex optimization problems. A repopulation scheme is employed, in which a set of elite individuals (that are also diverse among each other) are selected in the evolutionary process carried out by the EA. A multi-encoding strategy and a criterion that measures uniqueness among individuals are also proposed in order to drive the evolutionary process by favoring diversification or intensification. The proposed strategies were implemented in an EA and applied to find optimal schedules of power network outages. Tests were performed in different benchmark power networks and the obtained results reveal that the proposed strategies can help to obtain high-quality solutions in reduced computational times. This feature is particularly useful when solving many real problems, like the SNO, in which different suboptimal solutions are acceptable but one of them should be obtained in a reduced computational time.

References

1. Aickelin U, Bull L (2002) Partnering strategies for fitness evaluation in a pyramidal evolutionary algorithm. In: GECCO 2002: Proceedings of the genetic and evolutionary computation conference, New York, USA, 9–13 July 2002, pp 263–270
2. Aytug H, Koehler GJ (2000) New stopping criterion for genetic algorithms. Eur J Oper Res 126(3):662–674. doi:10.1016/S0377-2217(99)00321-5. http://www.sciencedirect.com/science/article/pii/S0377221799003215
3. Blum C, Roli A (2003) Metaheuristics in combinatorial optimization: Overview and conceptual comparison. ACM Comput Surv 35(3):268–308. http://doi.acm.org/10.1145/937503.937505
4. Chakraborty UK, Janikow CZ (2003) An analysis of gray versus binary encoding in genetic search. Inf Sci 156(3–4):253–269. doi:10.1016/S0020-0255(03)00178-6. http://www.sciencedirect.com/science/article/pii/S0020025503001786. Evolutionary Computation
5. Christie R (1999) Power systems test case archive. http://www.ee.washington.edu/research/pstca/
6. Crainic T, Toulouse M (2010) Parallel meta-heuristics. In: Gendreau M, Potvin JY (eds) Handbook of metaheuristics. International series in operations research and management science, vol 146. Springer US, pp 497–541. http://dx.doi.org/10.1007/978-1-4419-1665-5_17
7. Davidor Y (1990) Epistasis variance: suitability of a representation to genetic algorithms. Complex Syst 4(4):369–383
8. de Souza J, Brown Do Coutto Filho M, Ramos Roberto M (2011) A genetic-based methodology for evaluating requested outages of power network elements. IEEE Trans Power Syst, 26(4):2442–2449. doi:10.1109/TPWRS.2011.2138727
9. Deb K, Agrawal R (1995) Simulated binary crossover for continuous search space. Complex Syst 9(2):115–148
10. Droste S, Schmitt LM (2004) Feasible approaches to convergence results for evolutionary algorithms. Part ii: Runtime analysis of evolutionary algorithms and summary. Technical report, The University of Aizu, Japan
11. Eshelman LJ (1991) The chc adaptive search algorithm: how to have safe search when engaging in nontraditional genetic recombination. In: Rawlins GJE (ed) Foundations of genetic algorithms. Morgan Kaufmann, San Francisco, CA, pp 265–283. http://www.mpi-sb.mpg.de/services/library/proceedings/contents/foga90.html
12. Garey MR, Johnson DS (1979) Computers and intractability: a guide to the theory of NP-completeness. W. H. Freeman & Co., New York
13. Goldberg DE (1989) Genetic algorithms in search, optimization and machine learning, 1st edn. Addison-Wesley Longman Publishing Co Inc, Boston
14. Gouvêa MM, Araújo AFR (2015) Evolutionary algorithm with diversity-reference adaptive control in dynamic environments. Int J Artif Intell Tools 24(01):1450, 013–1–36. http://www.worldscientific.com/doi/abs/10.1142/S0218213014500134
15. Greenhalgh D, Marshall S (2000) Convergence criteria for genetic algorithms. SIAM J Comput 30(1):269–282. http://dx.doi.org/10.1137/S009753979732565X

16. Grefenstette JJ (1992) Genetic algorithms for changing environments. In: Männer R, Manderick B (eds) Parallel problem solving from nature 2, PPSN-II, Brussels, Belgium, September 28–30, 1992. Elsevier Science Inc., New York, pp 139–146

17. Hartl DL, Clark AG (2006) Principles of population genetics, 4th edn. Sinauer Associates, Inc

18. Holland JH (1992) Adaptation in natural and artificial systems. MIT Press, Cambridge

19. Jong KAD (2006) Evolutionary computation - a unified approach. MIT Press, Cambridge

20. Mathias K, Whitley L (1994) Changing representations during search: a comparative study of delta coding. Evol Comput 2(3):249–278. doi:10.1162/evco.1994.2.3.249

21. ONS: Planning outages of electrical network equipments - Brazilian iso network procedures report - in portuguese. Technical report, ONS, Rio de Janeiro, Brazil (2005)

22. Pendharkar PC, Koehler GJ (2007) A general steady state distribution based stopping criteria for finite length genetic algorithms. Eur J Oper Res 176(3):1436–1451. doi:10.1016/j.ejor. 2005.10.050

23. Pinedo ML (2012) Scheduling. Theory, algorithms, and systems, 4th edn. Springer, New York. doi:10.1007/978-1-4614-2361-4

24. Reeves C (2012) Genetic algorithms. In: Gendreau M, Potvin JY (eds) Handbook of metaheuristics. International series in operations research and management science, vol 146. Springer US, pp 109–139. http://dx.doi.org/10.1007/978-1-4419-1665-5_5

25. Rowe J, Whitley D, Barbulescu L, Watson JP (2004) Properties of gray and binary representations. Evol Comput 12(1):47–76. http://dx.doi.org/10.1162/evco.2004.12.1.47

26. Safe M, Carballido J, Ponzoni I, Brignole N (2004) On stopping criteria for genetic algorithms. Advances in artificial intelligence – SBIA 2004: Proceedings of the 17th Brazilian symposium on artificial intelligence, Sao Luis, Maranhao, Brazil, September 29–October 1, 2004. Springer, Berlin, pp 405–413. http://dx.doi.org/10.1007/978-3-540-28645-5_41

27. Shi L, Rasheed K (2012) A survey of fitness approximation methods applied in evolutionary algorithms. Computational intelligence in expensive optimization problems. Springer, Berlin, pp 3–28. http://dx.doi.org/10.1007/978-3-642-10701-6_1

28. Slatkin M (1985) Gene flow in natural populations. Annu Rev Ecol Syst 16:393–430. http://www.jstor.org/stable/2097054

29. Ursem RK (2002) Diversity-guided evolutionary algorithms. Parallel problem solving from nature—PPSN VII: Proceedings of the 7th international conference Granada, Spain, September 7–11, 2002. Springer, Berlin, pp 462–471. http://dx.doi.org/10.1007/3-540-45712-7_45

30. Whitley LD (1991) Fundamental principles of deception in genetic search. In: Rawlins GJE (ed) Foundations of genetic algorithms. Morgan Kaufmann, San Francisco, pp 221–241

31. Zanghi R, Roberto M, Souza J, Do Coutto Filho M (2012) Scheduling outages in transmission networks via genetic algorithms. In: 8th power plant and power system control symposium - IFAC proceedings volumes (IFAC-PapersOnline), vol 8. IFAC, pp 489–494. http://www.ifac-papersonline.net/Detailed/58429.html

Chapter 12
WBdetect: Particle Swarm Optimization for Segmenting Weld Beads in Radiographic Images

Rafael Miranda, Myriam Delgado, Tania Mezzadri, Ricardo Dutra da Silva, Marlon Vaz and Carla Marinho

The radiographic inspection of weld beads is important to ensure quality and safety in pipe networks. Visual fatigue, distractions, and the amount of radiographic images to be analyzed can be listed as main factors for human inspection errors. This chapter presents an approach for automatically segmenting weld beads in Double Wall Double Image (DWDI) X-ray photographs by combining two known methods in the literature: Particle Swarm Optimization (PSO) and Dynamic Time Warping (DTW). Vertical profiles of the weld beads are obtained from the windows' coordinates encoded by particles and compared, via DTW, with a predefined model. Experiments are performed considering two phases: first, tests are carried out to set the default configuration, and second the configured system (named WBdetect) is evaluated, including a comparison with another approach. Promising results show that WBdetect converges, most of the time, to the window that allows a proper segmentation of

R. Miranda (✉) · M. Delgado · T. Mezzadri · R. Dutra da Silva
Federal University of Technology of Paraná -UTFPR, Curitiba, Parana, Brazil
e-mail: arthur.rafa10@gmail.com

M. Delgado
e-mail: myriamdelg@utfpr.edu.br

T. Mezzadri
e-mail: mezzadri@utfpr.edu.br

R. Dutra da Silva
e-mail: rdutra@dainf.ct.utfpr.edu.br

M. Vaz
Federal Institute, Curitiba, Parana, Brazil
e-mail: marlonvaz@gmail.com

C. Marinho
CEN-PES/PDEP/TMEC - PETROBRÁS, Rio de Janeiro, Brazil
e-mail: carlamarinho@petrobras.com.br

© Springer International Publishing Switzerland 2017
N. Nedjah et al. (eds.), *Designing with Computational Intelligence*,
Studies in Computational Intelligence 664,
DOI 10.1007/978-3-319-44735-3_12

the weld bead, outperforming the compared approach (the average accuracy achieved by WBdetect is 93.63 +−12.91, and 65.88 +−17.9 % for the other approach).

12.1 Introduction

Petrochemical industries have networks of pipelines through which gases and liquids are transported. These networks of fluid conductive pipes are constructed by attaching pipes and other components by means of welded joints [17]. They are designed to tolerate great efforts and critical pressure and temperature conditions, since failure may cause serious damage to the environment, to the installations and processes. However, the welding process is subject to defects and flaws during the formation of the weld bead. Thus, in order to monitor the quality of weld beads, periodic inspections by means of nondestructive testing techniques (NDT) help to prevent such failures [1]. Radiography is a widely used method to inspect weld beads.

The inspection of weld beads is commonly performed by experts who, despite all training and knowledge, are prone to make mistakes due to eyestrain, knowledge level, number of images to be analyzed, fatigue, and distraction [8]. Because of these considerations, research centers have been focusing their efforts on the development of automatic or semiautomatic inspection systems for the interpretation of radiographic images of welds. Segmentation followed by defects classification in weld beads is an area that has attracted attention from the pattern recognition community. However, most approaches for classification of defects are associated with manual or semiautomatic segmentation of the welded joints. The reason is the challenging aspects of the weld bead in the images, such as orientation, size, and form [4]. Many techniques are not totally automatic and, usually, too specific, i.e., they are focused on the simplest cases (Single Wall Single Image (SWSI) or Double Wall Single Image (DWSI) radiographic images), and generally they can not handle Double Wall Double Image (DWDI).

Such difficulties as well as the need of an effective method for automatic segmentation of weld beads have motivated the proposition of the approach described in this work. This chapter is an extension of the paper published in [10]. The approach proposed in [10] also applies Particle Swarm Optimization (PSO) to DWDI radiographic images and uses Dynamic Time Warping (DTW) as a fitness function that measures the similarity between an ideal profile model and a profile extracted from the image. In this work the same methodology is adopted to design the proposed system named Weld Bead detector system (WBdetect) but the experiments have been extended to include a more consistent parameter setting phase, result analysis and comparison with another approach.

The association of PSO and DTW is incipient in the literature [11], and this work contributes in this area of research. PSO has been chosen since it is easy to implement and widely used for continuous optimization. Furthermore, PSO is able to combine random components and historical knowledge to guide itself through the search space, increasing the chances of the solution to be optimal [13, 18]. On the other hand, DTW is flexible to size and offset differences between two series being compared [3]. DTW can also be used to identify corresponding points between two series [2].

This chapter is structured as follows. In Sect. 12.2, the text describes the main problem being addressed. Section 12.3 provides an overview of the concepts used in the proposed approach. Section 12.4 presents the WBdetect: an approach to automatically detect weld beads in oil pipes based on particle swarm optimization and dynamic time warping. Experiments and results are described in Sect. 12.5. Finally, Sect. 12.6 concludes the chapter and provides directions for future works.

12.2 The Addressed Problem

The DWDI radiographic technique uses a source of X-rays positioned outside the pipe, at the location of the welded joint, and the radiation beam goes through the two walls of the pipe (Fig. 12.1). This technique is used when it is not possible to place the source or the film inside. The radiation source can be aligned to the weld, creating an image with overlapping arcs, or inclined (as in Fig. 12.1), producing a weld bead with elliptical shape on the radiographic film [1].

All the images considered in this work assume weld beads with the elliptical shape depicted in Fig. 12.1c. Those images are provided by the Center for Research and Development of Petrobrás (Centro de Pesquisas e Desenvolvimento da Petrobrás - CENPES) and represent operating conditions with different levels of light, noise, contrast, nonstandard formats, and dimensions in the weld beads.

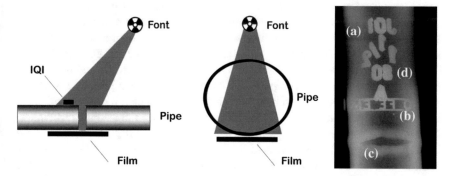

Fig. 12.1 Radiographic image of type DWDI. *Left* DWDI technique; *right* DWDI result emphasizing **a** the pipe wall; **b** the image quality indicator; **c** the weld bead; **d** the landmark indicating the exposition angle of welded joint

12.3 Background

This section presents the basic concepts necessary to understand the work, including related works (described in Sect. 12.3.1), fundamentals of Particle Swarm Optimization (Sect. 12.3.2) and Dynamic Time Warping (Sect. 12.3.3).

12.3.1 Related Works

This section describes related approaches to segment weld beads based on Computational Intelligence (mainly PSO). One of the first works is the segmentation method proposed in [9] which is based on the intensity of the pixels in the region of the weld bead. The method proved to be effective only if the weld bead appears in the image as a straight line. It was also observed that the intensity distribution of the weld bead along the lines of the image resembled to be Gaussian. In [4], the authors propose a method for extracting the weld bead in radiographic images using a Genetic Algorithm (GA). The method computes geometric transformations on a weld bead model looking for the best match with the weld bead in the image. It achieved good results for SWSI and DWSI, however, it did not have the same performance for the DWDI technique. A comparison between GA and PSO is presented in [8]. At first, the pipe is delimited and then the weld bead detection is performed by means of deformable masks. The approach obtained good results for DWSI and SWSI and for a specific subset DWDI images. Comparisons showed that PSO outperformed AG. In [5], a method to detect the central region of the weld bead in DWDI images is proposed. The method is composed of three steps: preprocessing, for noise reduction; PSO-based optimization, which searches for the best ellipse fitting the central region of the weld bead; and, finally, the selection of the best candidate region, considering similarity and focal length. A preliminary version of the proposed approach is presented in [10]. Some experiments were carried out and initial results were obtained. This chapter is an extension of that previous work [10] with the addition of new experiments, including the comparison with another approach [11]. The work presented in [11] served as an inspiration for [10] and this chapter. It can be divided into three stages: pipe identification (based on the thresholding and labeling techniques), detecting characters in the image which identify its quality (IQI) (this phase is also based on labeling but uses models of the IQI characters to identify the region to be eliminated from the original image), and finally the search for the weld bead that also combines PSO and DTW. After the detection and elimination of the tube and IQI characters, the method obtains the region of interest (ROI) which defines the search space of PSO particles. The next step is where [10, 11] mostly differ since in [11] the proposed approach was developed to find out only one arc at each time, so PSO must run twice for the complete segmentation of the weld bead. Further, each particle extracts a set of n profiles and the method calculates the similarity of these profiles (via DTW) with the synthetic model (that is a single Gaussian curve). Thus, with the DTW cost calculated for each one of the n profiles, the average is computed to measure the quality of the particle. Reference [11] represents a great contribution to our approach: (i) it served as an inspiration since it was the first work to joint PSO and DTW to detect weld beads in DWDI radiographic images; (ii) it provides the ROI in which our approach performs the search for the weld bead. In order to deal with some of the gaps left by previous works, we discuss in this chapter a new method (WBdetect) for segmentation of weld bead in DWDI radiographic images.

12.3.2 Particle Swarm Optimization

The PSO algorithm was developed by Kennedy and Eberhart in 1995 [7]. The inspiration came from the behavior of fish school and bird flock that use the strategy of collaborating to evolve. PSO became a widespread technique for its low computational cost and for allowing the share of information inherent to the social behavior of its individuals. The PSO algorithm randomly initializes its population over the search space and each particle $\mathbf{p}$ is associated with a velocity vector $\mathbf{v}$. The movement of the particles enables the search for better positions within the search space. The process is iterated until a stopping criterion is reached. The most commonly used criteria are: maximum number of iterations and acceptable fitness value [7]. The quality of each particle is measured by the fitness function. Particles store their experiences throughout evolution. The best position found by a particle is stored in personal best (**PBest**) and the best position found by the population is stored in global best (**GBest**) or local best (**LBest**) if a neighborhood structure is considered. These information are used to change the velocity that leads a particle toward a better fitness value [7, 16]. Equation 12.1 describes the velocity update performed at each iteration for each particle:

$$\mathbf{v}_i^{t+1} = w\mathbf{v}_i^t + C_1 r_1 (\mathbf{PBest}_i^t - \mathbf{p}_i^t) + C_2 r_2 (\mathbf{GBest}_i^t - \mathbf{p}_i^t), \tag{12.1}$$

such that $\mathbf{v}_i^t$ is the current velocity of the i-th particle at iteration t, w is the moment of inertia, which controls the search capability in particle space, $\mathbf{p}_i^t$ is the i-th candidate solution at the current iteration, r_1 and r_2 are random values uniformly distributed in the range $[0, 1]$ and C_1 and C_2 are two acceleration constants [13, 15, 16], which dictate the portion of local and global memory information that will contribute to the particle's movement. High values of w favor the previous move (global exploitation) while lower values favor a local search. The constants C_1 and C_2 also influence the search. Higher values diminish the influence of w. It is common to control that situation by initializing w with higher values which smoothly decay throughout iterations, promoting a balance between global and local searches. The particle coordinates are updated according to Eq. 12.2:

$$\mathbf{p}_i^{t+1} = \mathbf{p}_i^t + \mathbf{v}_i^{t+1}, \tag{12.2}$$

where $\mathbf{v}_i^{t+1}$ is the updated velocity and $\mathbf{p}_i^{t+1}$ is the updated position.

In this work, the position of each particle defines the coordinates of the window from which the weld bead profile will be extracted.

12.3.3 Dynamic Time Warping

The Dynamic Time Warping (DTW) algorithm searches for the best alignment between two series of values. The algorithm uses dynamic programming to measure the similarity between two series, minimizing the distance between them (commonly the Euclidean distance). It is used in areas such as speech, signature and gestures recognition, robotics, medicine, and data mining [2, 3, 14]. The use of DTW for image processing is still not widespread. One of the first applications to weld beads detection was presented in [11], work that has inspired the proposal presented in this chapter. DTW operation can be described as follows. Given two time series $A = \{a_1, a_2, \ldots, a_n\}$ and $B = \{b_1, b_2, \ldots, b_m\}$ of sizes n and m, the algorithm constructs a $n \times m$ matrix such that each element stores the distance $d(i, j)$ between points a_i and b_j, corresponding to the alignment of the points of the two series. A *path* is a contiguous set of elements of the matrix that defines the mapping between the series A and B (Fig. 12.2). The path is subject to some restrictions [3, 6, 14], it must begin and end at diagonally opposite corners of the matrix (limit restriction); it must not have leaps, ensuring the alignment does not omit important features (continuity constraint); it must not go back, ensuring the features do not repeat (monotonicity constraint).

The total cost of the path (alignment) is given by:

$$Cost(A, B) = \frac{1}{N} \sum_{i=1}^{K} e_i \qquad (12.3)$$

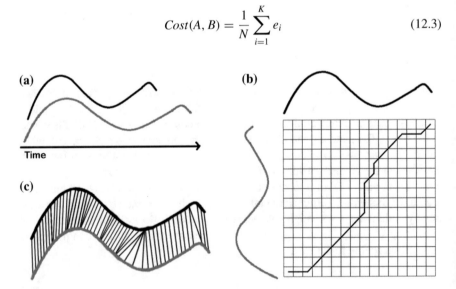

Fig. 12.2 Example of DTW. **a** Two series to be analyzed. **b** The distance matrix and the best path are emphasized. **c** Resulting alignment. Font: adapted from [6]

such that K is the number of elements in the path with elements e_i, and $N = n + m$. Out of many possible paths, the one that minimizes the cost function (3) is chosen [14]. The path can be found with dynamic programming according to the following steps [14]:

1. Initial condition:

$$D(1, 1) = 2d(1, 1);$$ (12.4)

2. Path equation:

$$D(i, j) = min[D(i, j - 1) + d(i, j), D(i - 1, j - 1) + 2d(i, j), D(i - 1, j) + d(i, j)];$$ (12.5)

3. Adjustment window:

$$|i - j| <= r,$$ (12.6)

where r is a positive integer called adjustment window, i and j are the current indexes of the matrix D. The adjustment window is used to speed up the running of DTW by limiting the data that is analyzed.

12.4 WBdetect: The Proposed Method

In this section we present the proposed approach named WBdetect. It consists of a PSO-based technique developed to automatically detect weld bead in radiographic images of oil pipes, which is associated with DTW to calculate the particles's fitness.

This work is part of a project developed by the Research Group on Images and Computer Vision (Grupo de Pesquisas em Imagens e Visão Computacional – GIVIC) of the Federal University of Technology, whose objective is to detect defects in weld beads by using DWDI radiographic images. For this purpose, the work developed in [11] performs a preprocessing to recognize the pipe, removes landmarks in the image, and retrieves a region of interest (ROI) where possibly is the weld bead (see Fig. 12.4, including the shaded area).

Our method performs a search in the ROI through a PSO algorithm with a DTW cost-based fitness, which returns the similarity between a profile extracted by a particle and a predefined model. WBdetect comprises three main steps to segment the weld bead: the construction of the synthetic model, the optimization performed by PSO and, finally, the fine-tuning process. The PSO optimization has two main steps: the search and extraction of a profile by the particle, and the measuring of similarity by means of DTW.

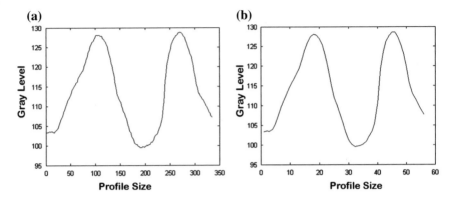

Fig. 12.3 Example of the adopted data reduction process: **a** before reduction, **b** after reduction

12.4.1 Synthetic Model

The distribution of intensities of pixels in a weld bead region of a DWDI image resembles the one shown in Fig. 12.1c, whose profile has two intensity peaks, corresponding to the bead arches, and one intensity valley, corresponding to the central region of the weld bead. A profile example (similar to the model) is shown in Fig. 12.3. The model is constructed by using a Gaussian distribution of gray levels [0, 255]. Initially, the model was defined with a complete set of points (Fig. 12.3a), as the proposal progressed, a reduction (Fig. 12.3b) was necessary based on a profile reduction factor.

An average of every Z points of the original profile is computed to represent the original points. The technique besides reducing the computational time, smooths noise in the profile, as illustrated in Fig. 12.3.

12.4.2 PSO Optimization

One of the first steps to design WBdetect is to define the search space limits of PSO particles. Due to the high processing time, it was observed the ROI needed to be reduced by 50% in its size (i.e., number of pixels). The search space was also reduced by eliminating the boarders of the image: a total of 20% of the image width is removed from each side (highlighted region in Fig. 12.4), since the intensity of the pixels in these regions is very homogeneous and would not resemble the pattern generated. Thus only 60% of the image width is used as a valid search space. The image height (HI) is not changed and the entire ROI height is considered a valid search space. The search for the candidate region that mostly resembles the weld bead model is performed by PSO, which randomly initializes the particles in the valid search space. The location of the window to extract the weld bead profile associated to the i-th particle at iteration t is given by the particle position $\mathbf{p}_i^t$.

Assuming a general particle position given by $\mathbf{p} = (p_1, p_2, p_3, p_4)$, the location of the associated window is defined based on the particle elements, where each element is defined as:

- p_1: position on the X axis of the profile extraction window;
- p_2: position on the Y axis of the profile extraction window;
- p_3: height H of window (the size of profile to be extracted);
- p_4: width L of window (number of pixels considered, where $L/2$ are on the left and $L/2$ are on the right side of the window central coordinate).

The values of p_1 and p_4 can be assumed in the valid search space (i.e., 60% of the ROI width), p_2 can assume values on the image height range (i.e., $p_2 \in [0, HI]$, remembering HI is the image height (ROI)). Theoretically p_3 could vary in $[0, HI]$ but we decided to reduce this range to $[H_{min}, H_{max}]$. Two possible ranges are going to be tested $[0.3, 0.5] * HI$ and $[0.2, 0.35] * HI$. The values of p_1, p_2, p_3 and p_4 are randomly generated within allowable limits. Particles that go beyond the search space limits are set to the first valid position. After all iterations of PSO, the values of p_1, p_2, p_3, and p_4 of **GBest** define the region of the candidate weld bead.

Vertical Profile Extraction

As previously discussed, the location of the window used to extract the weld bead profile is based on the particles encoding. An illustration of this association is depicted in Fig. 12.4. The left upper side of the window is defined by p_1 and p_2 while p_2 and p_3 define its final height. In this example, p_4 is defined as 1.

The profile computed using 1 pixel wide window (i.e., $p_4 = 1$) tends to be noisy (Figs. 12.4a and 12.5a). In order to avoid this problem, we impose a minimum value and the profile is computed using the average of $p_4 = L > L_{min}$ pixels for each line of a window centered at the vertical axis. Figure 12.4b illustrates (with $p_4 = L = 100$) the advantage of using larger windows.

The profile values are normalized in the range $[0, 255]$ depending on a threshold computed as the difference between the maximum and minimum gray levels. The normalization enhances the profile curves and its compatibility with the synthetic model.

Similarity Computation

Given the profile computed as described in the previous section and the synthetic model, the similarity between the series represented by the $A = \{a_1, a_2, \ldots, a_n\}$ and $B = \{b_1, b_2, \ldots, b_m\}$, can be computed based on the cost of DTW (Eq. 12.3). In this work, a version of the cost without normalization is also tested, according to Eq. 12.7, to reduce elongated profiles:

$$Cost(A, B) = \sum_{i=1}^{K} e_i. \tag{12.7}$$

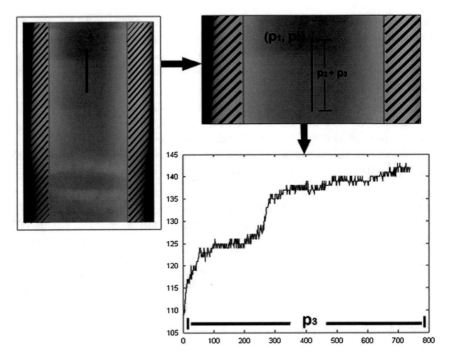

Fig. 12.4 Illustration of vertical profile extraction for an outside weld region

The reason for changing the computation of the cost of DTW is presented as follows. Assuming the cost of normalized DTW as $Cost_1$, when the value of p_3 (profile size) is increased, the new cost value, $Cost_2$, tends to be lower even if there is no target image resulting from this increment. This occurs mostly because the increased part of the pipe has a small contribution to the sum, but the cost decreases as the value of N increases - (see Eq. 12.3). Thus, an increase in p_3 usually produces a smaller DTW cost ($Cost_2 < Cost_1$) with no guarantee of improving the weld bead detection. As discussed in Sect. 12.4.1 we used in this work a technique to reduce the profile data, that, associated with the use of Eq. 12.7 can improve the segmentation capabilities of the method (see Sect. 12.5 for more details).

A perfect DTW alignment of two series would be a path corresponding to the diagonal of the distance matrix (see Fig. 12.2), but this perfect matching is so far to be found here. The optimal alignment is found by a very time consuming analysis of the whole matrix. In order to speed up the process, the search is performed inside a restricted diagonal window, since the farther the path is from the matrix diagonal the worse is the alignment. Our approach is based on the Sakoe–Chiba technique [14] and the value for the adjustment window is defined empirically. This work uses a modification to the technique proposed in [12] to ensure that the endpoints belong to the path. Another technique used to improve the computational cost of DTW was the inclusion of a lookup table that stores computations that have already been

(a)   **(b)**

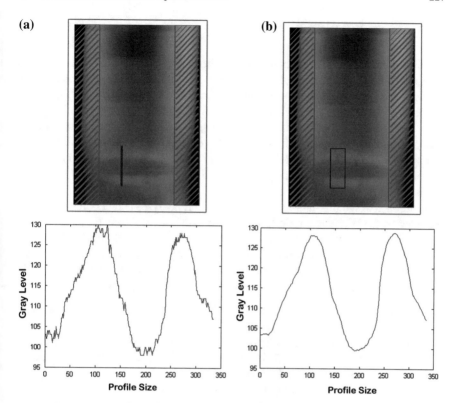

Fig. 12.5 Profile extraction. **a** simple profile extraction **b** average-based profile extraction

performed. The DTW cost and coordinates of a particle (integer coordinates in an image location) are stored so that if a particle goes to a previously visited location (a common behavior when PSO is almost converging to a solution and particles are close to each other) the fitness is not recomputed.

Fine-Tuning

A fine-tuning is performed after the PSO search if the output image requires adjustments (an extra part of the pipe is present or part of the weld bead is missing). Therefore, the region detected by a particle can be automatically incremented or decremented as an attempt to find out higher similarities with the synthetic profile model. The maximum amount of increment or decrement can be limited according to the height of the detected profile. All possibilities (discrete steps of increment or decrement) are performed and the one with the highest similarity (computed by DTW) to the model is chosen. If no improvement is obtained, the profile found originally by PSO is maintained.

By following the processes described in the previous sections the weld bead can be segmented. Next section discussed the results obtained by the application of WBdetect to a set of operating condition images.

12.5 Experiments and Results

In this section, the proposed approach is applied on a set of 30 DWDI radiographic images of welded joints. From the total of available images, 10 are selected to set the configuration parameters and the remaining 20 are used to test the proposed approach. For each image, PSO is executed 30 times (each round with a different random seed). After each PSO run, the fine-tuning process can take place. In this work, the limit of change (increase or decrease) in fine-tuning is chosen as a percentage (30%) of the original profile size returned by PSO. A total of 6 possible increments and 6 possible decrements is considered, ranging from 10 to 30% in the lower or upper part of the profile.

At the end of the fine-tuning process, the expert analyzes and classifies the result provided by the WBdetect, as follows. Similar to the methodology proposed in [10], some numerical criteria (grades) are established to differentiate the results. According to the type of information that could be provided by the proposed approach to the GIVIC system (i.e., the system that detects defect in weld beads, which is out of the scope of this chapter), five quality levels are considered by the expert:

- in the case of a total error (i.e., the weld bead is not detected) the result can not be accepted and the output receives grade 0 (zero);
- when the output contains only a portion (one arc of the weld bead - either the top or bottom) it receives grade 25, indicating that even with a missing arc, this image could be used as an input to the GIVIC system;
- if part of one arc is missing (part of top or part of bottom) it receives grade 50, since we still have missing parts but we have more information than the previous case;
- when the weld bead is segmented, but there is a great part of image with no target, the result receives grade 75, indicating that now we have the total information necessary to weld bead defect detection but the chance of false positives is higher than that of correct segmentation;
- finally, when the weld bead is segmented properly it receives the maximum grade 100.

This results classification analysis is performed by the expert (for all the images and rounds considered - in this work for example, a total of $30*30 = 900$ images must be analyzed for each approach). Some examples of final grades (one for each case) are illustrated in Fig. 12.6.

In order to analyze the performance of the proposed approach considering different parameters to be set by the user, experiments have been carried out aiming to define the default configuration. Next section details those experiments (named configuration phase).

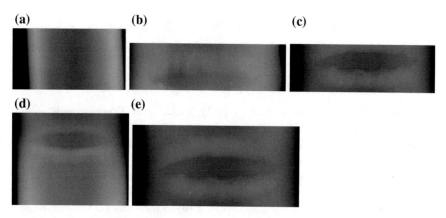

Fig. 12.6 Examples of the proposed classification scheme for the weld bead detection: **a** unacceptable (grade 0); **b** partially acceptable (grade 25); **c** regular (grade 50); **d** acceptable (grade 75); **e** totally correct (grade 100)

12.5.1 Setting Parameters of WBdetect

The following parameters are considered in the configuration phase: the swarm size S; the inertia moment w, and acceleration constants C_1 and C_2 (Eq. 12.1); *maximum* of *iterations* performed by PSO; the *range* R_{p_3} in which parameter p_3 can vary (it is measured proportional to the ROI height (HI)); the width L of the window provided by each particle (this parameter is encoded in the element p_4 of the particle); the *adjustment window* r (Eq. 12.6) adopted in DTW to limit the computational time; the form of computing DTW cost which can be *normalized* or not (Eqs. 12.3 and 12.7, respectively) and finally the *reduction* factor imposed to profile data in order to reduce the processing time. Table 12.1 shows the parameter values considered during the configuration phase.

Due to the low budget available for computational time, a small swarm ($S = 15$) and a larger value for C_2 compared with C_1 were adopted as an attempt to speed up the PSO convergence.

Table 12.1 Tested parameters

PSO							DTW		Profile
S	w	C_1	C_2	p_4 (L)	R_{p_3}	Iter (max)	Window	Norm	Reduction factor
15	[0.9, 0.4]	0.7	1.2	100	$[0.2, 0.35]*HI$	{25, 35, 40}	{30, 50}	yes/no	{6, 10}
					$[0.3, 0.5]*HI$				
					Evolve $\in [0, HI]$				

Assuming a fixed subset of configuration parameters given by

- **FixSubConfig**: {$S = 15$, w initialized as 0.9 and gradually decreased until it achieves 0.4, $C_1 = 0.7$, $C_2 = 1.2$ and $p_4 = L = 100$},

the sets of alternative configurations considered in this phase are:

- **ConfigA**: Fix + {max iter = 35; $R_{p_3} = [0.2, 0.35]*HI$; $r = 50$; no normalization in DTW; reduction factor = 6;}
- **ConfigB**: Fix + {max iter = 35; $R_{p_3} = [0.3, 0.5]*HI$; $r = 50$; no normalization in DTW; reduction factor = 6}, (i.e., ConfigA except for R_{p_3});
- **ConfigC**: Fix + {max iter = 35; $R_{p_3} = [0.3, 0.5]*HI$; $r = 50$; normalization in DTW; reduction factor = 6}, (i.e., ConfigB except for DTW normalization);
- **ConfigD**: Fix + {max iter = 40; p_3 evolve $\in [0, HI]$; $r = 50$; no normalization in DTW; reduction factor = 6}, (i.e., ConfigA except for max iter and R_{p_3});
- **ConfigE**: Fix + {max iter = 40; p_3 evolve $\in [0, HI]$; $r = 30$; no normalization in DTW; reduction factor = 10}, (i.e., ConfigD except for reduction factor and DTW adjustment window r);
- **ConfigF**: Fix + {max iter = 25; p_3 evolve $\in [0, HI]$; $r = 50$; no normalization in DTW; reduction factor = 6}, (i.e., ConfigD except for max iter);
- **ConfigG**: Fix + {max iter = 25; p_3 evolve $\in [0, HI]$; $r = 30$; no normalization in DTW; reduction factor = 10}, (i.e., ConfigD except for max iter, reduction factor and DTW adjustment window r).

Results provided by each configuration on a set of 10 images are illustrated in Table 12.2.

Configurations D to G are attempts to improve the flexibility of the model by evolving the height of the profile extraction window. However, they do not provide competitive results. To decide if there is significant difference among configurations, we applied the Friedman statistical test with the level of significance $\alpha = 5\%$. According to the Friedman analysis (we obtained p-value = 0.0010) there is difference in at least one pair of configurations. We applied a post hoc analysis (based on Tukey–Kramer test) to verify which pairs are different from each other. Figure 12.7 shows that configuration A (solid line—blue) is better than configurations E–G and

Table 12.2 Results of the configuration phase

	Accuracy (%) average	StDev (%)	Average time (s)
ConfigA	**93.40**	10.38	28.47
ConfigB	89.38	10.11	26.52
ConfigC	81.39	9.89	23.78
ConfigD	82.15	10.82	16.72
ConfigE	80.83	14.17	8.48
ConfigF	80.77	11.36	12.19
ConfigG	82.64	10.19	**6.44**

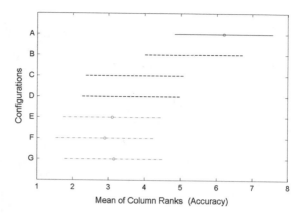

Fig. 12.7 Post hoc analysis of Friedman test for the different configurations

performs equal to configurations B-D. Although the lack of statistical difference in the performances of configurations A, B, C, D, the WBdetect default configuration is defined as ConfigA, due to its higher accuracy average.

In the next section, we present some results of WBdetect (i.e., the proposed approach with ConfigA) applied to the set of 20 remaining images.

12.5.2   WBdetect Results

When applied to the test set, WBdetect achieved an accuracy rate of $93.63 +-12.91\%$, demonstrating its capacity to perform well even for non trained patterns. With the default configuration, WBdetect is capable of reducing the cost along iterations as depicted in Fig. 12.8, which shows the result of one round for one specific image among the set of 20 images and 30 rounds each.

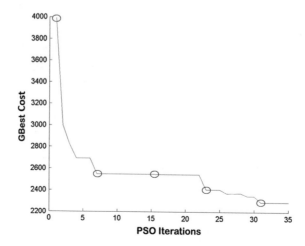

Fig. 12.8 One example (among 30 rounds) of GBest evolution along iterations with emphasis at some specific points

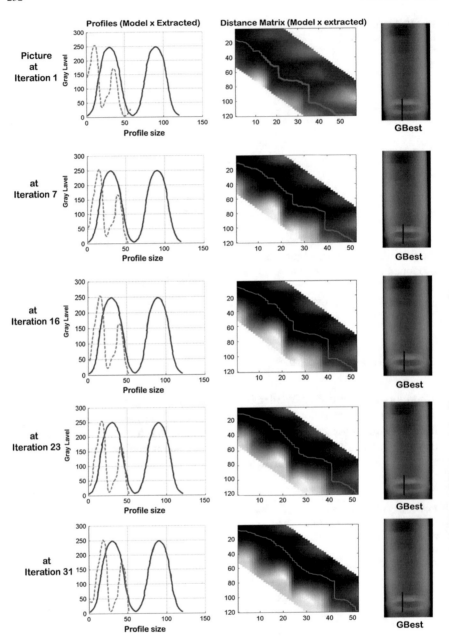

Fig. 12.9 Details of WBdetect information taken among evolution (for those points emphasized in Fig. 12.7)

From the evolutionary process depicted in Fig. 12.8, some specific stages are chosen to be scrutinized. Figure 12.9 illustrates the details resulting from the solution provided by the best particle in the swarm at different stages of the evolutionary process (e.g., at iterations 1, 7, 16, 23, 31). On the left side of Fig. 12.9, we have the comparison between the profile obtained by the particle with that defined by the synthetic model. On the right side we have the window location provided by the best particle. In the middle we have the distance matrix resulting from this matching, which is used during the DTW cost computation (i.e., the fitness computation). We can notice that the best *path* in the distance matrix goes toward diagonal (the perfect matching between profiles) along iterations. This process is accompanied by improvements of the coordinates of profile window as well as in the profiles matching (specially for the first part of the compared profiles which represents the half top of the weld bead).

Table 12.3 Results for the test set

Image	WBdetect		Vaz (2015)	
	Accuracy (%)	Time (s)	Accuracy (%)	Time (s)
1	94.17	12.87	26.67	57.5
2	96.67	20.7	69.17	54.77
3	75.00	25.5	70.00	53.77
4	100.00	16.17	80.83	54.4
5	100.00	16.53	76.67	56.7
6	100.00	19.33	75.83	57.37
7	95.00	25.77	72.50	56.4
8	50.00	16.67	63.33	64.33
9	100.00	15.77	69.17	57.33
10	89.17	18.53	43.33	57.8
11	100.00	14.83	40.00	58.43
12	100.00	18.5	10.00	58.03
13	99.17	24.23	75.00	59.93
14	100.00	33.8	72.50	54.47
15	100.00	40.27	90.00	53.97
16	100.00	30.77	75.00	57.27
17	74.17	31.1	77.50	55.13
18	100.00	43.07	80.00	58
19	100.00	44.1	75.00	58.2
20	99.17	47.5	75.00	54.1
Average (among images)	**93.63**	**25.8**	65.88	56.9
StDev	12.91	10.47	17.84	2.5

12.5.3   Comparison with the Literature

This section performs a comparison between WBdetect and the approach described in [11] since it is the most related approach and one of the few works that also addresses DWDI radiographic images. To perform a fair comparison, the same expert analyzes the results of both approaches on the same set of 20 test images. Table 12.3 shows the results of this comparison (computed over 30 runs of PSO) for each image, indicating that WBdetect has a highly better performance when compared with the approach proposed by [11] and besides it runs faster (practically takes half of the time).

The superior performance of WBdetect in terms of accuracy can be explained because it considers the whole weld bead in each particle. In [11] the weld bead detection is decoupled for the top and bottom arcs which can provide inconsistent results.

12.6   Conclusion

This chapter detailed WBdetect, an automatic weld bead detection approach, as part of a support system for the inspection of welded joints in pipes used in the oil industry. The main goal of the proposed approach is to combine two techniques already known in the literature, PSO and DTW, both applied in a challenging problem in the pattern recognition area: segmentation of weld beads in DWDI radiographic images. Although the two techniques are quite known, their joint use is little explored in the literature. The use of PSO can be justified by the fact that it is widely adopted in continuous optimization (but achieved good results even for discrete optimization) and uses information and historical knowledge to guide the search, in addition to requiring low computational cost. DTW is suitable to handle the comparison between series of different sizes.

The method for weld bead segmentation discussed in this chapter could be summarized in three main steps: the design of synthetic model, the optimization performed by the PSO particles, and the fine-tuning process. After the synthetic model phase, PSO-based optimization (the most important step) was responsible for extracting the image profile. The calculation of the quality of each solution was based on a comparison with the synthetic profile model and took into account the cost function calculated by DTW. The fine-tuning process was used as an attempt to find a better similarity profile after the end of PSO search.

Experiments were divided in three parts: (i) configuration phase, (ii) exploring the best configuration; (iii) comparing WBdetect with another approach. The configuration tests considered a set of alternative parameters but two of them affected more the results: (ia) the range of the height (encoded in p_3 particle element) of the profile extraction window and (ib) how to compute the particle's fitness. In the first case, we considered two main assumptions: ia1) element p_3 of every particle is randomly initialized into the range R_{p_3} and remains fixed along evolution; ia2) element p_3 evolves along evolution. In the second case, the fitness could be calculated in two different forms: (ib1) based on a normalized version of the DTW cost, or ib2) based on a non-normalized version of the cost. The configuration with nonadaptive height (p_3 does not evolve) associated with non-normalized version of the fitness computation achieved the best results in terms of average accuracy during the training phase. This configuration was chosen to set the standard proposed approach named WBdetect.

WBdetect is capable to find out acceptable solutions with a reduced computational cost memory (few particles in a few iterations). This occurred partly because the observed behavior regarding the limits of the search space and the profile size helped in reducing the PSO search space. The proposed amendments to the DTW allowed faster execution, reducing the average running time from 15 min (in a very preliminary version of WBdetect) to a maximum of 30 s per image (the first set of configurations A-C) and up to 17 s (for the second group D-G).

The main contribution of this work focuses on a little explored area in the literature: the weld bead segmentation on DWDI radiographic images. In general, most of approaches are semiautomatic or can be too specific (they can not be applied to DWDI). An important aspect to be noted regarding the application context is that few weld radiograph inspection works use images derived from actual operating conditions with different levels of light, noise, contrast, nonstandard formats and dimensions as the ones used in this work. It is valid to point out that the images used in this work are challenging for most of standard solutions adopted in the literature, that generally work well with uniform images, but are of little practical use in real ones. The inspection of oil pipeline radiographic images is a subjective task and only a qualified inspector can provide a technical report. Thus, the proposed approach can support these professionals by automatically segmenting weld beads to which subsequent defect detection can be performed.

For future work we intend to resume the version with the adaptive profile window height aiming to evolve p_3. One of the possibilities is to define a penalty if the fitness exceeds a threshold height. However, other modifications can be handled in order to achieve more competitive results, since the inclusion of this parameter allows greater flexibility to the system. We also intend to perform a comparison between other existing ways to calculate the path in DTW. Finally, we aim to expand the image database and implement other DWDI weld bead detection techniques that can serve as a comparison for this work, since the currently available references are more specific to other types of images.

Acknowledgments This work has been partially supported by the Brazilian National Research Council (Conselho Nacional de Pesquisa - CNPq), under research grant 309197/2014-7 to M.R. Delgado. The authors would also like to thank Leopoldo Americo Miguez de Mello R&D Center - CENPES, Brazilian Petroleum - PETROBRÁS.

References

1. Andreucci R, Radiologia industrial (2014). http://www.abendi.org.br/abendi/Upload/file
   /Radiologia-Jul-2014.pdf
2. de Azevedo ACG (2011) Implementação de um método para reconhecimento on-line de assinat-
   uras. http://bdm.unb.br/bitstream/10483/1538/1/2011_AugustoCesarGoncalvesdeAzevedo.
   pdf
3. Chino DYT (2014) Mineração de padrões frequentes em séries temporais para
   apoio à tomada de decisão em agrometereologia. Master's thesis, USP - São Car-
   los, http://www.teses.usp.br/teses/disponiveis/55/55134/tde-04062014-142915/publico
   /dissertacaoRevisadaChino.pdf
4. Felisberto MK, Lopes HS, Centeno TM, De Arruda LVR (2006) An object detection and
   recognition system for weld bead extraction from digital radiographs. Comput Vis Image
   Underst 102(3):238–249
5. Suyama FM, Krefer AG, Faria AR, Centeno TM (2013) Identificação da região central de
   cordões de solda em imagens radiográficas de tubulações de petróleo do tipo parede dupla
   vista dupla. In: Proceedings of 4th Brazilian conference on intelligent systems (BRACIS), pp
   1–11. Sociedade Brasileira de Computação
6. Jeske J (2011) Similaridade de séries temporais na bolsa de valores. http://www.lume.ufrgs.
   br/bitstream/handle/10183/31034/000782115.pdf?sequence=1
7. Kennedy J, Eberhart R (1995) Particle swarm optimization. In: Proceedings of the IEEE International Conference on Neural Networks, vol 4, pp 1942–1948
8. Kroetz MG (2012) Sistema de apoio na inspeção radiográfica computadorizada de juntas sol-
   dadas de tubulações de petróleo. Master's thesis, CPGEI - Federal University of Technology
9. Liao TW, Ni J (1996) An automated radiographic ndt system for weld inspection: part i - weld
   extraction. Ndt E Int 29(3):157–162
10. Miranda RAR, Delgado MR, Centeno TM, da Silva RD, de Oliveira Vaz, M (2015) Enxame
    de Partículas para segmentação de cordão de solda em imagens radiográficas. In: Anais do
    XII Congresso Brasileiro de Inteligência Computacional, pp 1–6. Sociedade Brasileira de
    Inteligência Computacional
11. de Oliveira Vaz M (2015) Detecção de cordão de solda em imagens radi-
    ográficas do tipo parede dupla vista dupla através do PSO e do DTW. Tech.
    rep., CPGEI - Federal University of Technology, Curitiba, Paraná, Brazil.
    http://www.dainf.ct.utfpr.edu.br/~myriam/Technical_Report_Download/Relatorio_Tecnico
    _PSO_DTW_MarlonVaz.pdf
12. Paliwal KK, Agarwal A, Sinha SS (1982) A modification over sakoe and chiba's dynamic time
    warping algorithm for isolated word recognition. Signal Process 4(4):329–333
13. Perlin HA, Lopes HS, Centeno TM (2008) Particle swarm optimization for object recognition
    in computer vision. In: New frontiers in applied artificial intelligence, pp 11–21. Springer,
    Heidelberg
14. Sakoe H, Chiba S (1978) Dynamic programming algorithm optimization for spoken word
    recognition. IEEE Trans Acoust Speech Signal Process 26(1):43–49
15. Shi Y, Eberhart R (1998) A modified particle swarm optimizer. In: The 1998 IEEE interna-
    tional conference on evolutionary computation proceedings, 1998. IEEE world congress on
    computational intelligence, pp 69–73. IEEE, New York
16. Shi Y, Eberhart RC (1999) Empirical study of particle swarm optimization. In: Proceedings of
    the 1999 congress on evolutionary computation, CEC 99, vol 3. IEEE, New York
17. Telles PCdS (1976) Tubulações industriais. In: Tubulações industriais. Livros Técnicos e Cien-
    tíficos
18. Waintraub M, Pereira C, Schirru R (2009) Parallel particle swarm optimization algorithm in
    nuclear problems. http://www.iaea.org/inis/collection/NCLCollectionStore

Author Index

© Springer International Publishing Switzerland 2017
N. Nedjah et al. (eds.), *Designing with Computational Intelligence*,
Studies in Computational Intelligence 664,
DOI 10.1007/978-3-319-44735-3

237

Printed in the United States
By Bookmasters